GEOLOGY IN SHROPSHIRE

DEDICATION
This book is dedicated to the memory of Philla Davis.

GEOLOGY IN SHROPSHIRE

Peter Toghill

Acknowledgements

My sincere thanks go to Gill Jenkinson who has taken so much trouble in the accurate and clear drawing of all the fifty-four text figures. Joan Stockdale has typed the manuscript, often with many versions of particular sections, and I am most grateful to her for her patience. Dr. Peter Osborne has kindly read and commented on Chapter 11, the 'Ice Age', for me, and Dr. Colin Dixon has allowed me to use his unpublished information on the geology of the Breidden Hills. Mr. O'Hare and Mr. Thursby of Grinshill Stone Quarries have given me information on the present use of Grinshill Stone and allowed me to visit their quarries at Grinshill and Myddle. The photograph of the Condover mammoth site (Plate 30) was taken by Mrs. Eve Roberts and she has allowed me to use it. The Editor of the journal *Field Studies* has given permission for me to reproduce (with amendments) figures from Toghill and Chell, 1984, Shropshire Geology, *Field Studies*, Vol. 6, pp. 59-101. The figures in this book are Figs. 4, 5, 6, 10, 11, 16, 18, 19, 22, 23, 26, 29. I am grateful to Chris Musson who has allowed me to use his aerial photograph of the Longmynd Scarp Fault (Plate 11).

Cover shows the ancient volcanic hills of Caer Caradoc and The Lawley looking across the Church Stretton Fault Valley from the Longmynd.

Copyright © Peter Toghill, 1990

British Library Cataloguing in Publication Data available

ISBN 1 85310 090 0

This edition first published in the UK 1990 by Swan Hill Press, an imprint of Airlife Publishing Ltd.

All rights reserved. No part of this book may be reproduced or transmitted in any form or by any means, electronic or mechanical including photocopying, recording or by any information storage and retrieval system, without permission from the Publisher in writing.

Swan Hill Press
An Imprint of Airlife Publishing
101 Longden Road, Shrewsbury SY3 9EB, England.

Contents

	Introduction	
Chapter 1	The Fundamentals of Geology	1
Chapter 2	Using Shropshire to understand Geology	9
Chapter 3	Shropshire's Oldest Rocks	17
Chapter 4	The first signs of abundant life – the Cambrian period in Shropshire	43
Chapter 5	Volcanoes, Deep and Shallow Seas – the Ordovician Period	58
Chapter 6	Coral Reefs and the end of the Iapetus Ocean – the Silurian Period	88
Chapter 7	All Shropshire above sea level – the Devonian period	107
Chapter 8	Shallow Seas, Deltas and Coal Swamps – the Carboniferous Period	115
Chapter 9	Deserts and Salt Lakes – the Permian and Triassic periods	137
Chapter 10	Pieces of the Jigsaw missing – the Jurassic, Cretaceous and Tertiary Periods	155
Chapter 11	Towards the Present Day – the Quaternary Period, including the Ice Age	162
	Selected References	181
	Index	184

About the author

Dr. Peter Toghill has lived in Church Stretton since 1971 and is a Lecturer in Geology in the School of Continuing Studies (previously the Extramural Department) at Birmingham University. He is based at The Gateway, Shrewsbury, from where he also organises the University's adult education courses in all subjects in Shropshire. He is well known in Shrewsbury and Shropshire for his lecture courses on all aspects of geology. He read Geology at Birmingham University and received his doctorate in 1965. Before coming to Shropshire he worked as a government geologist for the British Geological Survey, and the Natural History Museum in London. A Fellow of the Geological Society of London he has always had an interest in the conservation of geological sites for teaching and was secretary of the Geological Society of London's Conservation Committee from 1980 to 1985.

In 1979 he initiated the formation of the now thriving Shropshire Geological Society and is now its Vice-President. He is also a member of the Ironbridge Gorge Museum's Geology Committee.

About the book

The geology of Shropshire is more varied than any other area of comparable size in Britain, and perhaps the world. Eleven of the thirteen recognised geological periods of time are represented by rocks in the county, ranging in age from the ancient metamorphic rocks and volcanic lavas of the Wrekin area around 700 to 600 million years old, to the glacial sediments of the last Ice Age only 13,000 years old which contained the famous Condover mammoths. The variety of Shropshire's rocks and fossils has attracted geologists since the early nineteenth century, including many eminent academics and countless numbers of students. Nowadays the general public is also becoming more and more interested in geology and there is also an ever growing number of knowledgeable amateurs.

This book is intended to appeal to those interested in geology at any level. It is an authoritative account of the Geology of Shropshire, and the County's unique rock sequence is also used to explained many of the general principles of geology. The book contains 54 text figures and 32 photographs; including the most common fossils of the various geological periods, and also includes the only available up-to-date geological map of the whole of Shropshire.

Introduction

No other area of comparable size in Britain displays such a variety of geology as Shropshire, with rocks representative of eleven of the thirteen recognised periods of geological time, ranging in age from about 700 million years old to those formed in the last Ice Age, a few thousand years ago. The many different rock types have given rise to the great variety of scenery which the county exhibits — the gaunt Stiperstones, the Wrekin, the Longmynd plateau, the famous escarpment of Wenlock Edge, the great plain of North Shropshire, to name just a few features.

Geology, as an all embracing study of the origin, structure, composition and history of the Earth from its formation around 4,600 million years ago to the present day, is a subject which is becoming more and more appreciated by the general public. Natural disasters, earthquakes and volcanic eruptions, are nowadays explained in the newspapers and on television using technical terms which would never have been used twenty years ago, such as plates, faults and subduction zones. Western society's dependence on fossil fuels has also brought geology into everyone's household. New fossil finds such as the Condover mammoths, found in September 1986, are headline news.

Shropshire saw many of the great pioneers of British geology in Victorian times, and today it still attracts countless students of geology and physical geography at all levels, from schools to universities, and more so today the interested amateur who has perhaps learned his geology through a university adult education evening class. The Shropshire Geological Society, founded in 1979, drew its initial support from the latter group. Because of its classic place in British geology, Shropshire is an excellent area to learn about some of the general principles of geology, as well as learning the details of the county's own unique rock sequence and its development through 700 million years of geological time, and this is what this book sets out to explain.

Conservation and Access to Sites

Although this account is not an itinerary of geological sites in Shropshire, it does of course mention and describe a number of classic sites, the majority of which are on private land. No site should ever be visited without the permission of the owner and the onus for obtaining this rests entirely with the reader. The conservation of geological sites has attracted a good deal of attention in the 1970s and 1980s, particularly in areas such as Shropshire which receives so many geological parties. Everyone should follow the Country Code, and there are also Codes of Conduct for geological parties and geologists produced by the Geologists' Association and the Institution of Geologists. These can be obtained from both these bodies at Burlington House, Piccadilly, London, W1.

Leaders of parties should never allow indiscriminate hammering and fossils should only be collected from loose scree.

Chapter 1
The Fundamentals of Geology

One of the most difficult things for the layman to grasp is the vast scale of geological time. One day in 1986 I went to look at a fossil which had been found in a stream flowing through a garden near Church Stretton. When I arrived and examined it, it turned out to be a beautiful coral washed down from the limestone of Wenlock Edge during the melting of the last Ice Age. The owners were deeply interested in what I told them, and particularly when I said the fossil was 420 million years-old. I'm sure they believed me, even if they were very surprised. Can the rocks and fossils of Wenlock Edge really be 420 million years old, and were they really formed in a warm, shallow, subtropical sea, near to the equator, with coral reefs growing as in the Caribbean today?

We geologists take all this for granted, although we constantly argue amongst ourselves about the detailed ages of particular rocks, as will be seen later. I usually start to explain geological time by asking people to accept that 2 cm (about one inch) could be worn away from the Longmynd in Shropshire, by rain, wind and snow, in their lifetime without their noticing anything. 2 cm in 100 years works out as 200 m (660 feet) in 1 million years, or 8,000 metres (26,000 feet) in 40 million years. Therefore, in this time a mountain range as high as the Himalayas could be completely worn away, or conversely could be raised up to that height. The Himalayas are in fact about 40 million years-old, as are the Alps, and so we can come to accept the length of geological time if we accept that everything happens slowly but surely over vast intervals of time.

Modern calculations on the South Downs suggest they are being eroded at a rate of about one inch in 500 years, but rates of erosion will vary greatly according to climate, and relative rates will depend on whether mountain ranges are still rising as the Himalayas are today.

If we accept that the Earth is incredibly old, about 4,600 million years, the other great dictum of geology is that the present is the key to the past — uniformitarianism, a fundamental principle first put forward by the Scottish gentleman farmer, James Hutton, in his book *Theory of the Earth* in 1795, and later summarized by the great Victorian geologist Charles Lyell in his *Principles of Geology* published in 1830. Lyell gave a copy of his book to Charles Darwin, and inspired him to think about the subject. This dictum says, quite simply, that we can explain everything we see in the ancient rocks of the Earth by a process happening somewhere on the Earth today — in other words, the processes happening on the Earth today are exactly those which operated in the past. The Earth 500 million years ago would have had oceans, volcanoes, deserts, and an atmosphere with weather systems similar to those of today. Although this principle is broadly true, certain things were different in the distant past. No plants or animals existed on land until about 400-350 million years ago. The atmosphere prior to 600 million years ago was less rich in oxygen than it is today, and prior to 4,000 million years ago the volcanic activity on the Earth was probably much more intense than it is today. However, in studying the broad principles of the formation of rocks, minerals and fossils, and the processes which cause volcanoes, earthquakes and continental drift, uniformitarianism is a very useful principle. Our exploration of the planets has added to this, since in looking at the present surface of the Moon we see features, the vast numbers of meteorite craters, which the Earth almost certainly exhibited about 4,000 million years ago but which have long since been destroyed by processes of earth movements and erosion which have never taken place on the Moon.

So using uniformitarionism we can look at the rocks of Wenlock Edge, 420 million years-old, and say that these look very much like the coral reef areas of the present Caribbean. The fossil reefs within the Wenlock Limestone occur as patches with a fat discus shape, just like the patch reefs around the Bahamas. These reefs today only form in warm, shallow, clear subtropical seas, and so the area around Wenlock Edge was like this 420 million years ago. Does this mean that the British latitudes were subtropical then, or does it mean that Britain lay nearer to the equator? The now proven hypothesis of continental drift, embodied in the new concept of Plate Tectonics, explains this quite easily, and proves the latter answer. Southern Britain (England and Wales) was closer to the equator, and in fact throughout the last 600 million years has moved from latitudes around 60°S to the present position of 50°N to 60°N, having spent about 200 million years near to the equator. As the movement took place, southern Britain experienced various changes in climate, from temperate to tropical and back to temperate. The Ice Ages are world-wide climatic changes of short duration, not caused by the European continent drifting into, and out of, polar areas.

As in our calculations on vertical movements of mountain ranges, we can calculate that these horizontal continental drift movements were slow but sure: a movement of 100 degrees of latitude across the globe is about 12,000 km (7,000 miles) and if it took place in 600 million years, this is a rate of 20 km (12 miles) in one million years, or 1 km in 50,000 years. A little more division will bring the rate down to 2 cm per year, or about one inch per year. So in a human lifetime (100 years), Britain may have moved two metres or about eight feet north-eastwards, along with the whole of Europe. No one would really notice this, but over millions of years it would add up to very large movements.

These very small continental movements can be measured today using sophisticated equipment. In 1985 the distance between Sweden and Massachusetts (USA) was found to be widening by 1.4 cm per year. This was calculated using a number of telescopes at different places in the two countries simultaneously observing the same distant radio source in space, and using extremely accurate measurements of time. These measurements show the Atlantic Ocean is widening at this steady rate, and we now know the volcanic mid-Atlantic ridge is producing the new oceanic crust, in the form of lava flows, which causes the ocean to widen. Conversely, the same study showed the distance between Massachusetts and a station in Texas to be shrinking by 1 cm a year. North America is being compressed. Measurements in other oceans now show that some ocean floors are moving at up to 12 cm a year, and these horizontal movements are much quicker than the 2 cm in *100* years which we postulated earlier for the vertical movement of mountain ranges.

Geologists have always known that continents change their shapes, mountains are squeezed up and rift valleys open, but the process of the widening and closure of oceans, sea floor spreading, was only discovered in the 1960s. It was the final proof of a mechanism for continental drift and laid the foundations for the all embracing concept of plate tectonics, which can now be briefly explained.

The rigid outer layer of the Earth's surface down to about 100 km, called the lithosphere, is now known to be made up of a number of individual larger and smaller areas called plates, and plate tectonics (Fig. 1) shows how these move about, grow or decrease in size, collide and override each other, and occasionally slide past each other along major fault lines called transform faults, where great earthquake activity occurs, such as the San Andreas Fault in California (Fig. 2).

Sea flow spreading shows that oceans like the Atlantic and the Pacific are having new basaltic ocean crust continually produced along the line of

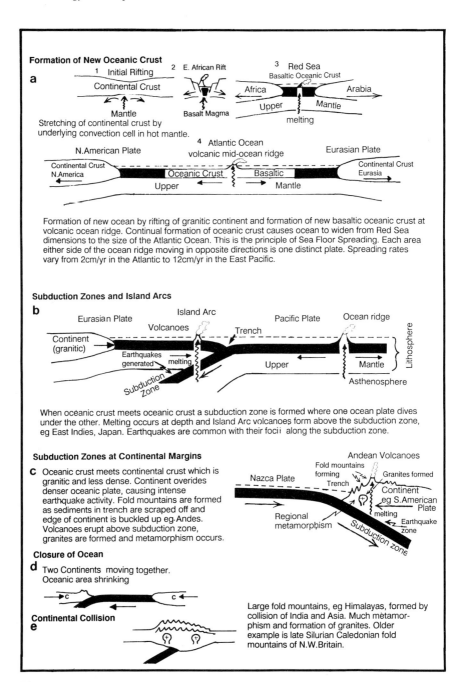

Fig. 1. The general principles of Plate Tectonics.

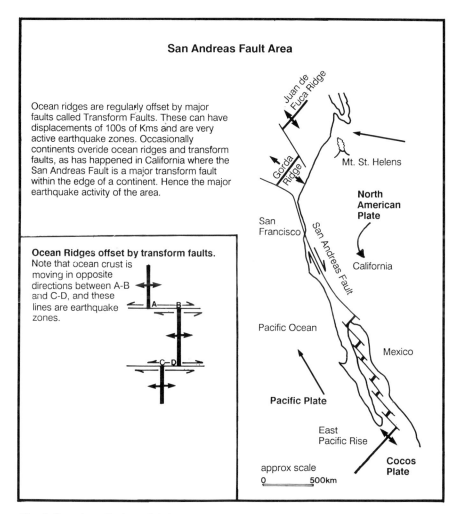

Fig. 2. Transform Faults and their role in Plate Tectonics.

volcanic ocean ridges such as the mid-Atlantic ridge. However, whereas the Atlantic is widening by about 2 cm a year, the Pacific is shrinking because all around its edges it is being overriden by lighter granitic continental crust, such as where the western side of South America is moving west and overriding the Nazca oceanic plate. All around the Pacific deep ocean trenches have formed as the oceanic crust dives down under the continental crust along what are called subduction zones. These subduction zones are the sites of widespread earthquake activity, volcanoes, metamorphism and mountain building (orogenesis) as shown in Fig. 1. Thus in the geological past where we see widespread volcanic lavas and ashes, particularly andesites and rhyolites, metamorphic rocks, and the presence of old mountain ranges, these prove the presence of ancient subduction zones.

Armed with these new facts, geologists have gone out in the 1970s and 1980s to show how Britain and the rest of the world have developed in the last 1,000 million years using plate tectonics concepts. Ancient volcanoes, deserts, deep seas, shallow subtropical seas, earthquakes, tropical rain forests — all these have affected Britain in the past due to our varying position on the globe throughout this time. At present we are in a temperate climate within a stable geological area, but there is no reason why in, say, fifty million years things could not change; indeed as I have explained above, they are bound to.

Dividing up geological time

Having said that the Earth is about 4,600 million years-old, it is sensible to divide up the vast amount of time into smaller divisions: eons, eras, and particularly geological periods (Table 1). Each geological period covers a number of millions of years, and during each period the rocks laid down are referred to as a system. The divisions are not arbitary, but have been based since their inception in the mid-1800s to around 1900, on major changes in the evolution of Life as found fossilised in the rocks of each geological system as it follows another in a definite sedimentary sequence. Thus the Silurian System of rocks laid down in the Silurian period, 435-405 million years ago, has a distinct set of fossils which distinguishes it from the Ordovician System below and the Devonian above, and this distinction is true throughout the world even though the rock types may be very different. The Silurian rocks of Australia look very different from those of Britain, but often contain the same fossils and so can be shown to be the same age. Systems are further divided into Series and Stages, and Periods into Epochs. The study of the layered sequence of rocks laid down as the various geological Systems is called Stratigraphy.

The identification of rocks of the same age by the fossils they contain was established by the surveyor William Smith in the early 1800s when he was working on the construction of the Thames-Avon canal system. He is often called the 'Father of English Geology', and in 1815 he published the first geological map of England and Wales. He firmly established the succession of rocks, and could be called Britain's first stratigrapher. Fig. 3 shows various fossils from different ages of rocks of the Ordovician period in Britain to show how the rocks can be subdivided and recognised in different parts of the country.

The divisions of geological time thus established produced a *relative* time scale by 1900. No one knew exactly how old the rocks were in *absolute*

The Fundamentals of Geology 7

Table 1. The Geological Time Scale:

EONS		ERAS	PERIODS/SYSTEMS		EPOCHS/SERIES	
PHANEROZOIC		CAINOZOIC	QUATERNARY		Holocene	10000
					Pleistocene	2
			NEOGENE	TERTIARY	Pliocene	
					Miocene	24
			PALAEOGENE		Oligocene Eocene Palaeocene	66
		MESOZOIC	CRETACEOUS		Upper Lower	135
			JURASSIC		Upper Middle Lower	205
			TRIASSIC		Upper Middle Lower	250
		PALAEOZOIC UPPER	PERMIAN		Upper Lower	290
			CARBONIFEROUS		Stephanian Westphalian Namurian Dinantian	355
			DEVONIAN		Upper Middle Lower	405
		PALAEOZOIC LOWER	SILURIAN		Pridoli Ludlow Wenlock Llandovery	435
			ORDOVICIAN		Ashgill Caradoc Llandeilo Llanvirn Arenig	492
			CAMBRIAN		Tremadoc Merioneth St. David's Comley	570
PRECAMBRIAN (CRYPTOZOIC)	PROTEROZOIC	LATE	900	PRECAMBRIAN	Dates in Millions of years ago	
		MIDDLE	1600			
		EARLY	2500			
	ARCHAEAN	LATE	3000			
		MIDDLE	3400			
		EARLY	3800			
	PRE-ARCHAEAN		4600 Myrs Formation of Earth			

8 *Geology in Shropshire*

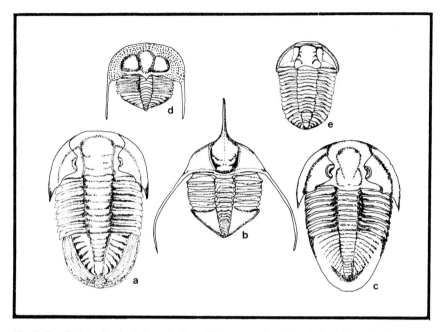

Fig. 3. Fossils (all trilobites) from the five different epochs of the Ordovician period to prove William Smith's dictum that rocks of different ages can be recognised in various parts of the country by the distinct fossils they contain: a. *Ogygiocaris selwyni*, Arenig Series, x0.7; b. *Ampyx linleyensis*, Llanvirn Series, x1; c. *Basilicus tyrannus*, Llandeilo Series, x0.4; d. *Onnia gracilis*, Caradoc Series, x1; e. *Flexicalymene quadrata*, Ashgill Series, x1.

terms, although we know now that the Ordovician period followed the Cambrian period, etc. The modern science of geochronology, pioneered by Professor Arthur Holmes in the early part of this century, dates igneous rocks by radiometric methods using the radioactive decay of certain elements present in minerals. This method has produced the absolute dates shown on Table 1. Even today, these figures are not completely accurate and are always being slightly modified.

Chapter 2
Using Shropshire to understand Geology

A glance at the geological map of Shropshire (Fig. 4) (after Toghill and Chell, 1984) shows the great variety of rock types of the different geological periods present in the county. Fig. 5 (after Toghill and Chell, 1984) shows the various folds and faults which affect the rocks, and is a structural map. Fig. 6 is a stratigraphical column and shows the various types of sedimentary rocks formed during geological time in Shropshire, and in particular whether they are marine, non-marine, or volcanic ashes and lavas. This diagram also shows the periods of major earth movements (orogenies) giving rise to folds and faults, and major fold mountains elsewhere. These would also be times of earthquake activity, together with the emplacement of igneous rocks at depth, the formation of metamorphic rocks, and the extrusion of volcanic lavas and ashes in many places in Britain, but not necessarily always in Shropshire.

All these diagrams show that Shropshire exhibits a more varied display of geology than any other area of comparable size in Britain, and possibly the world. We can thus use the county to explain a large number of aspects of geology: earthquakes, volcanoes, folds, faults, fossils, sedimentary and volcanic environments, which apply not just to Shropshire but to anywhere in Britain.

Having said all this, it is not surprising that the county saw many of the pioneers of early geology in Victorian times. Perhaps the most famous names associated with the county are: Murchison, who founded his Silurian System partly on the rocks of the Wenlock Edge and Ludlow areas; Lapworth, who founded the Ordovician system and worked on the rocks of that area around Shelve, together with Watts; and Cobbold, who investigated the Cambrian rocks of Shropshire in great detail. Many other names could be mentioned, but suffice it so say that even today numerous geologists still research the rocks of the county, and it is always a teaching ground for new generations of geologists.

A generalised account of Shropshire Geology

At this stage it is worth giving a brief account of the geology of Shropshire to provide an idea of what the county displays in terms of rocks and scenery.

In general terms, it is possible to say that the North Shropshire Plain, north of the Severn, and all areas east of the Severn, are made of relatively young (less than 300 million years-old) soft rocks, and form relatively low lying ground. This is in marked contrast to the South Shropshire Hills, and areas in the far north west, where the underlying rocks are much older (between 700 and 300 million years-old), and are much harder and resistant. However, even within the South Shropshire Hills the variety of the geology is such that each hill mass has its own distinctive feature, from the volcanic hog-backs of the Wrekin and Caer Caradoc, to the gaunt crags of the Stiperstones, the Longmynd plateau, and the marvellous ridge and vale country around Wenlock Edge.

Of the thirteen geological periods recognised, eleven are represented in Shropshire by different rocks and scenery. This is remarkable when one considers that in an area like Snowdonia only three geological periods are represented. The best way to describe the geology and scenery of the county is to consider each geological period in turn, from the oldest to the youngest, and to consider the rocks formed in that period, and the scenery they produce today.

We know little about the Earth's early history in Shropshire until we come to a time, within the late Precambrian, or Proterozoic, eon between 700 and 570 million years ago, when the oldest rocks in the county were formed. Very small patches of metamorphic rocks (Rushton Schists and Primrose Hill Gneisses) occur near to the Wrekin, and are about 700 million years-old, but the oldest widely occurring rocks are volcanic lavas and ashes, about 650-600 million years-old, forming the Wrekin, the Lawley and Caer Caradoc, and many other hills on the east side of the Church Stretton Valley, as well as Earl's Hill near Pontesbury. We don't know the location of the volcanoes that threw out all the lava and ash, and so although the Wrekin is made of volcanic rock, it never was a volcano. Later in this period, 590-575 million years ago, Shropshire lay under a shallow sea in which sandstones and shales were laid down which now form the Longmynd plateau. The deep valleys or 'batches' on the Longmynd lay bare the now vertical layers of rocks which were once horizontal on the sea floor. These ancient rocks contain no fossils since life had yet to appear on Earth in abundance. During this period a major break, or fault, appeared in the Earth's crust, which we call the Church Stretton Fault. This was to be a line of violent earthquakes for many millions of years but happily it has been dormant for about fifty million years.

During the Cambrian, Ordovician and Silurian periods, 570-405 million years ago, Shropshire lay almost continuously under the sea, shallow in the east and deeper in the west. Sometimes areas like the Longmynd and the Stiperstones stood out as islands. On the sea floor thicknesses of sandstones, shales (layered muds); and limestones accumulated, all with the entombed remains of animals living and dying on the sea floor. These remains would become the fossils we find today, the Cambrian period in Shropshire containing the earliest abundant fossils we find anywhere in Britain.

The rocks formed during these periods now form the Stiperstones-Shelve area, where the rocks are rich in lead ore and barytes; the volcanic rocks of the Breidden Hills; areas south east of the Wrekin; and the ridge and vale country stretching south east of the Church Stretton Hills, up to and including Wenlock Edge. In this latter area there is a wonderful demonstration of soft shales forming the valleys like Ape Dale and Hope Dale, and hard sandstones and limestones forming ridges like Hoar Edge and Wenlock Edge. Clun Forest, the Long Mountain and the Ludlow area are formed by Silurian rocks, the latter a classic area for the study of the Ludlow Series.

Wenlock Edge deserves special mention since it is perhaps the best example of escarpment in Britian, formed by a gently tilted layer of limestone, the Wenlock Limestone, with a steep westerly face and a gentle dip-slope to the east. The escarpment trends NE-SW and runs virtually unbroken for 30 km from Ironbridge to Craven Arms. A second layer of limestone forms a higher and parallel escarpment to the south east of Wenlock Edge itself with the valley of Hope Dale always between. The Wenlock Limestone was laid down under a warm shallow tropical sea in which coral reefs grew, now fossilised on Wenlock Edge. Britain lay over the equator at that time (the Silurian period), as we have explained earlier.

During the late Silurian and Devonian period, 405-355 million years ago, the whole of Shropshire rose above the sea for the first time. All rocks formed during this period were laid down in lakes and shallow lagoons as red and green sandstones, marls and limestones. We call these rocks the Old Red Sandstone and they form all of Corve Dale and the Clee Hills, although the highest parts of the Clees are of Carboniferous age.

Again, most of Shropshire was covered by the sea during the Carboniferous period, 355-290 million years ago. Only small patches of rocks

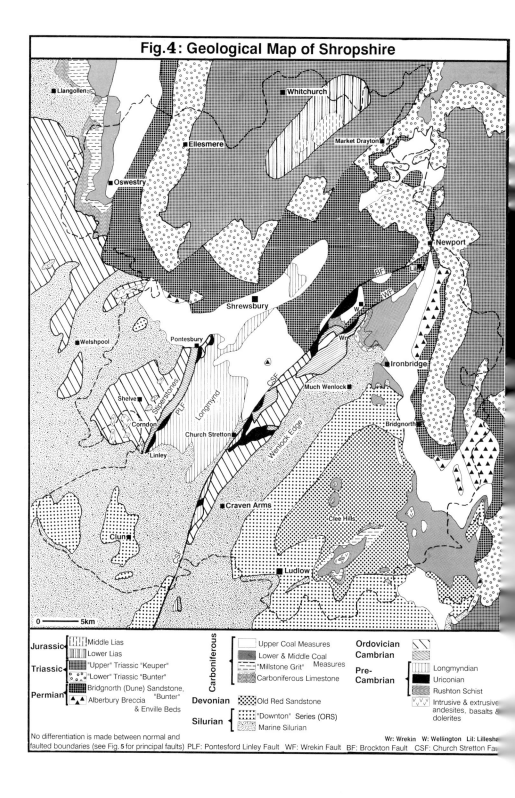

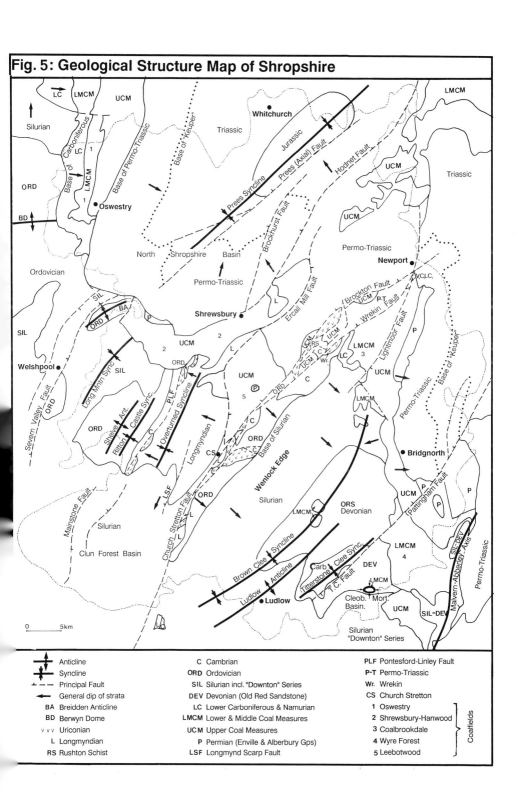

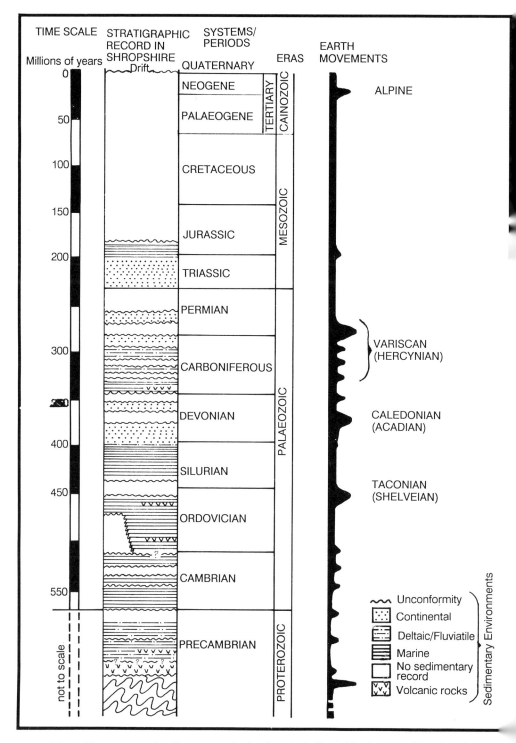

Fig. 6. The sequence of sedimentary and volcanic rocks in Shropshire (stratigraphic column) and main periods of earth movements. Intrusive igneous rocks not shown. (After Toghill and Chell, 1984.)

formed during the earlier part of the period; the Carboniferous Limestone and Millstone Grit (which are the dominant rocks of the Pennines) occur in south Shropshire; on the Clee Hills and the south east flanks of the Wrekin. Quite large areas of Carboniferous Limestone and 'Millstone Grit' occur north west of Llanymynech Hill northwards.

Everywhere in Britain the period is famous for its coal deposits, which were formed by the decay of vegetation in a tropical climate, in widespread deltas invading the shallow seas. While these occur in the north west of the county, and the top of the Clee Hills, the main area for deposits is around Ironbridge and Telford. It was here that coal, with local iron ore and limestone, cradled the Industrial Revolution. Little coal is nowadays obtained in Shropshire but some opencast mining yields coal and fireclay; and of course in the past imported Cornish ceramic clays were used to produce fine porcelain at Coalport. Less rich coal deposits occur in the southern part of the North Shropshire Plain, in what is known as the Shrewsbury-Hanwood coalfield, now abandoned. Igneous rocks (dolerite or dhustone) of that period form the summits of the Clee Hills, the highest in the county.

Nearly all the North Shropshire Plain and areas east of the Severn, particularly around Bridgnorth, are formed of bright red, reddish brown and pale brown sandstones of the Permian and Triassic periods, 290-205 million years-old. They form isolated hills such as Grinshill, Nesscliffe and Hawkstone in the north, and elsewhere rolling pastoral country. It is hard to believe that these rocks were formed in a desert area just like the Sahara today, with Britain still close to the equator but entirely a land mass. The pale brown Grinshill Stone is Shropshire's most well known (and best) building stone, and the only one to be exported from the county on a large scale.

The youngest consolidated, or solid rocks, as geologists call them, in the county are limestones and shales of the early Jurassic period, 200 million years-old, around Prees in north Shropshire. Laid down under a shallow sea which covered the earlier 'Sahara desert' they contain abundant marine fossils. This patch of rock is really a little piece of the rocks which form the Cotswolds left behind in Shropshire.

We now come to a period where little is known of what was happening in Shropshire. No rocks belonging to the late Jurassic, Cretaceous and Tertiary periods (190-2 million years ago) are found in the county, and so, for instance, we have no Chalk Downs which formed during the Cretaceous period in southern England.

About two million years ago the climate in Britain became much colder and we entered the great Ice Age of the Quaternary period. Ice sheets of the Pleistocene epoch covered Shropshire in three, four, or possibly many more successive advances from the north and west. There was certainly a glacier in the Church Stretton valley as recently as 20,000 years ago, and Mammoths roamed around Condover 13,000 years ago. When the ice sheets melted they left behind, particularly on the North Shropshire Plain, a thick cover of clays, sand and gravel, which we call Glacial Drift. This, over large areas, covers the solid rocks, and gives rise to rich agricultural land. The Ice Age left us with two important features in the county. The north Shropshire Meres and Mosses, and the River Severn. This used to flow northwards from Welshpool to the Dee estuary, but when this outflow was blocked by ice sheets in the north, it formed a series of lakes which eventually overflowed eastwards and cut the Ironbridge Gorge, and then the river flowed southward through Bridgnorth to join the Avon drainage. When the ice melted in the north it did not revert to its old course.

It should now be clear that the variety of the county's rock scenery reflects the countless events which have affected its geological evolution. We can now look in more detail at the various geological periods, their rocks, minerals, fossils and geological structures.

Chapter 3
Shropshire's Oldest Rocks

The oldest rocks in Shropshire belong to the Precambrian. This vast interval of time lasted from the origin of the Earth 4,600 million years ago to the appearance of abundant fossilised life forms, with hard exoskeletons, at the start of the Cambrian period, 570 million years ago. It cannot be conventionally divided into smaller geological periods because of the lack of abundant fossils. In recent times the Precambrian has been divided into subdivisions of quite large time intervals based on cycles of earth movements, and there is also a three-fold division into the pre-Archaean, Archaean and Proterozoic eons, the latter time interval (meaning former life) being represented by rocks in Shropshire. The term Precambrian is that commonly used to cover all rocks formed before the Cambrian. The Shropshire Precambrian rocks were formed at the very end of the Proterozoic between about 700 and 570 million years ago. A small patch of rocks around the village of Rushton, just west of the Wrekin (Fig. 4) represents the oldest in the county and could be around 700 million years-old. These very old Rushton Schists are metamorphic rocks (rocks altered by heat and pressure) and the other Precambrian rocks are the Uriconian Volcanic rocks, 650-600 million years-old, and the Longmyndian Supergroup of sedimentary rocks, 590-575 million years-old. There are also very small patches of metamorphic rocks at the south west end of the Wrekin, the Primrose Hill Gneisses and Schists which are difficult to interpret.

The Rushton Schists and Primrose Hill Gneisses and Schists

Schists and Gneisses are high grade metamorphic rocks formed by the alteration of pre-existing rocks by heat and pressure, and contain new minerals formed from the original chemical composition. If the heat and pressure are applied over large areas, for instance, in areas of orogeny and mountain building, the phenomenon is called regional metamorphism. If the effects are more localised by the presence of, say, a hot granite intrusion, this is called thermal or contact metamorphism. When sedimentary rocks, such as shales (layered muds now hardened with obvious bedding planes), are regionally metamorphosed they first change to fine grained slates with a very obvious splitting direction called the cleavage. With increased temperature and pressure, the grain size grows, new minerals appear, the cleavage becomes less uniform and a banding or foliation appears in the rock, usually with mica and garnet present. The rock is now a schist and all the minerals are aligned in one direction, the direction of least stress. Eventually, with more heat and pressure, it will turn into a gneiss, a coarse-grained rock with a very obvious colour banded appearance caused by segregation of new minerals along preferred orientation planes. Garnet-mica schists are common rocks in the Highlands of Scotland, and the Lewisian Gneiss in north west Scotland at over 2,000 million years-old is Britain's oldest rock type.

Anyone seeing the Rushton Schist for the first time will not be impressed with it when compared with Scottish Schists, nor does it have an immediate metamorphic appearance. Banding and foliation is just about visible in hand specimens, but under the microscope the metamorphic minerals are apparent. The rock is poorly exposed in a few sunken tracks and stream sections around and north west of Rushton over an area, which is mainly covered by glacial drift, of two square kilometres. It is a greenish brown, fine-grained rock with a poor foliation. In thin section under the microscope, quartz, mica, epidote and fresh garnet are visible. It is a metamorphic rock derived from a shale sandstone sequence of older rocks. So this, at about 700 million years-old, is Shropshire's oldest rock! As a schist it does not compare with those of Scotland, but it is probably a fragment of the older basement underlying the English Midlands.

The Primrose Hill Gneisses and Schists are even more puzzling to the layman. At the south west end of the Wrekin is a small hill marked on maps as Little Hill, but also known in earlier times as Primrose Hill. This little 'bump' in the ridge profile of the Wrekin is well seen from the north-west or south-east (Plate 2). The rocks in question only cover an area of 300 square metres. The gneisses are coarse grained, derived from igneous rocks, and the schists include green hornblende schists. The whole group is cut by veins of pink granite material known as granophyre. The rocks show similarities with the metamorphic rocks of the Malvern Hills and again could be part of the Precambrian basement, underlying the Uriconian Volcanics and Longmyndian sediments now to be described.

The Uriconian Volcanics

The Precambrian volcanic rocks of Shropshire do not present such problems of identification as the older metamorphic rocks. The volcanic hills of Shropshire, as exemplified by the Wrekin (Plates 1 and 2), form a spectacular and unique feature of the Shropshire landscape. Very late Precambrian volcanism is well known from many parts of southern Britain and can be explained using the plate tectonics concept. During the late Precambrian, about 700 million years ago, England and Wales, together with south-east Newfoundland were attached to Africa, India and the rest of Alfred Wegener's Gondwanaland, and lay 60-70° south of the equator. This latitude is estimated from palaeomagnetic studies, which shows the position of ancient land masses with relation to the magnetic poles. Scotland and Northern Ireland, as well as north-west Newfoundland, Scandinavia, Greenland and North America lay thousands of kilometres further north, near to the equator. The position of all these continental masses will be discussed in more detail in the next chapter, when we consider the Iapetus Ocean.

However, returning to Precambrian volcanism over southern Britain, a subduction zone formed about 650 million years ago on the 'northern' side of Gondwana, in the region of England and Wales, whereby continental

Plate 1. The Wrekin from the south west showing its misleading 'volcano like' profile.

20 *Geology in Shropshire*

Plate 2. The Wrekin from the north-west showing a 'hog back' ridge. The small hill at the south west (right hand) end of the ridge is Little or Primrose Hill.

crust began to override oceanic crust with a subduction zone near to what is now Anglesey (Fig. 7). On the continental side of this subduction zone volcanic activity broke out in Shropshire and elsewhere, just as it does today on the western side of both North and South American continental plates. The eruption of Mount St. Helens in the USA in 1980, and of Nevado del Ruiz in Columbia in 1986, are recent examples of major eruptions of this type (Fig. 7).

The Shropshire volcanic activity occurring between about 650 and 600 million years ago gave rise to lava flows and volcanic ashes from volcanoes long since eroded away, so that although we say the Wrekin (Plates 1 and 2) is made of volcanic rock, we cannot find any evidence of a vent, so we cannot say the Wrekin is an extinct volcano, nor Caer Caradoc and its associated hills.

The Uriconian Volcanic rocks are found now in two distinct belts running north east-south west (Figs. 4 and 5) and form conspicious hog back volcanic hills. However, in late Precambrian times, around 630 million years ago, the whole of Wales, Shropshire and the English Midlands was probably a continental area with volcanoes scattered about. What we see now are the erosional remnants following millions of years of earth movements and erosion. The two belts of volcanic rock follow the

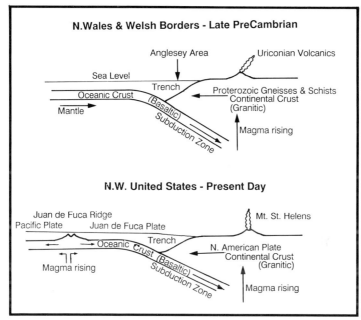

Fig. 7. Plate tectonics explanation for the formation of the Uriconian Volcanics in the late Precambrian, and a present day equivalent situation at Mount Saint Helens.

lines of two major faults (breaks in the Earth's crust). One of these, the Church Stretton Fault, is particularly well known and it is appropriate to discuss it now.

The Church Stretton Fault and associated faults

Major faults are breaks in the Earth's crust associated with regular earthquake activity, such as the famous San Andreas Fault in California. Rocks either side of a fault move with respect to each other either vertically or horizontally (Fig. 8). Minor stresses within the Earth can cause very small faults to develop and which only have a small displacement, maybe only a few centimetres, or up to a few metres. Many rock sequences show faulting of this type. However, if an area of the Earth is situated in one of the mobile belts with continuous but slow earth movements, usually at plate boundaries, e.g. Circum Pacific and Mediterranean areas at the present time, then once a fault is initiated, further earthquakes along it can make the total movement either side of it add up to many kilometres over millions of years. Indeed, areas west of the San Andreas Fault have already moved 560 km northwards in the last 150 million years (105 km in the last six million years) with respect to the areas

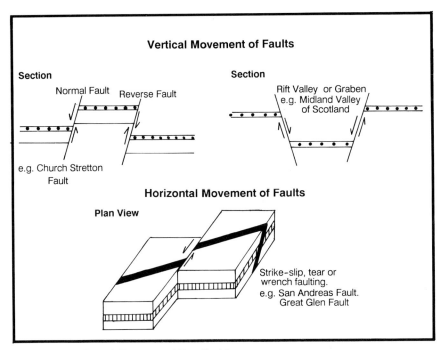

Fig. 8. Faults. Normal, Reverse and Tear Faulting.

on the east side of the fault (Fig. 2), and the area is still in motion. The whole process is caused because rocks near to the surface are brittle, and when enormous stresses are applied to them they eventually break.

As continents move over millions of years from areas of intense earth movements with volcanoes and earthquakes — the mobile belts, to more stable areas, the major fault lines become inactive. In Britain our most famous fault is probably the Great Glen Fault, which is followed by that dramatic defile which contains Loch Ness. Although it has been inactive for over 300 million years, the movement either side amounted to a total horizontal displacement of 100 km (or possibly 200 km) when it was active in Devonian times. So literally, north west Scotland has moved south west by 100 km with respect to Scotland south east of the Great Glen. Going back to our idea of slow but continuous movement, if this fault moved 0.5 m in 100 years, that is, a major earthquake every century, then it would take twenty million years for the 100 km displacement to occur. This is a possible life span for a major fault, although some would be active for much longer, as we shall see for the Church Stretton Fault. The Great Glen Fault was probably active for 100 million years, which implies a slower movement than 0.5 m in 100 years, or possibly a sparodic history including long periods of inactivity.

Shropshire now lies on the western margin of the stable Eurasian plate. Minor earthquakes such as that of 16 July 1984 are only minor readjustments along old fault lines, in this case a fault near Portmadoc in North Wales, probably caused by stresses set up by the continual widening of the Atlantic Ocean. However things were different in the past. During the late Precambrian, as well as volcanic activity, two major faults appeared in the Shropshire area and were active here, and further afield, as earthquake belts for many millions of years. The Church Stretton Fault (Figs. 4 and 5) was active about 650 million years ago until as recently as only fifty million years ago, and thus in its heyday would have rivalled the great San Andreas Fault in the intensity of its earthquakes. All this would have been caused initially by the subduction zone under Anglesey, referred to earlier, which created the stresses and formed the fault. It was most active during the Precambrian to Silurian periods 650 to 400 million years ago.

The Church Stretton Fault can now be traced from South Wales to near Newport in Shropshire, but it is the dramatic Church Stretton valley following the line of the fault which has given it its name, and also in this region the fault's effects were the most pronounced. It is difficult to say how much movement there has been along the fault, since it has not always moved in the same direction, sometime it has been a horizontal tear fault, and towards the end of its life the movement was entirely vertical as a normal fault. During the Tertiary, about fifty million years ago, its final movement was a vertical movement of 1,000 m, a downward displacement (downthrow) on its western side causing Wenlock Limestone just like that on Wenlock Edge 6 km to the south-east to be dropped down into the Church Stretton valley (Fig. 19).

Although the fault was initiated at the same time as the Uriconian Volcanics started to erupt, it is unlikely that much, if any, volcanic material came up the fault line. The eruptions were restricted to volcanic cones or fissures scattered about Shropshire and southern Britain. So it was a barren continent, no plant life or animal life on land had evolved, and any life in the sea would have been mainly marine algae. The land was probably reddish due to the first significant oxygen in the atmosphere giving rise to red oxidized rocks, and the volcanoes continually erupted, and earthquakes continued along the Church Stretton Fault. A short distance to the north-west, another important fault was initiated at the same time, the Pontesford-Linley Fault (Figs. 4 and 5), traced now from Pontesford near Pontesbury, just north west of the Longmynd, south east for 18 km to Linley. Further south west the fault is traceable into Wales as a disturbance rather than a fault. This fault was to continue later to play a vital part, along with the Church Stretton Fault, in the evolution of the Shropshire rock sequence, but ceased to be active at an earlier date, about 350 million years ago.

The detailed sequence of the Uriconian Volcanics

The above rocks now only occur along the line of the Church Stretton Fault and, to a lesser extent, the Pontesford-Linley Fault. Starting from the north east end of the Church Stretton Fault, the first conspicuous outcrop is at Lilleshall Hill (Fig. 4), with its well known column, and continuing south west we reach the famous area of the Ercall and the Wrekin. This area is what geologists refer to as the type area of the Uriconian Volcanics, in that they were first studied here, and are well exposed. The Wrekin is composed of alternations of pink and purple lavas known as rhyolites, which were very sticky when erupted, and ashes (called tuffs when found as consolidated rocks), some of which are rhyolitic and some andestic. Andesite is a lava named after the Andes where large amounts are erupted today. Volcanic ashes have the same chemical composition as the lavas of the same name since they are often formed by lavas solidifying in the vent and being blown out by explosions. So an andestic ash is derived from an andesite lava. Occasionally ashes may contain fragments of other rocks which the vent is penetrating. Ash can be in the form of coarse fragments which rain down close to the volcano, often containing large lumps of lava which hardened in flight, known as volcanic bombs, or in the form of the finest dust clouds which rise into the upper atmosphere and stay there for many weeks, affecting the sunsets, before falling to earth many miles away. Material of intermediate grain size can fall many kilometres away from the volcano as well. Some very fine ash bands called ignimbrites, welded tuffs or ash flows, are deposited from glowing gas cloud eruptions known as *nuée ardentes*. These clouds have temperatures in them in excess of 1,000°C and when they sweep down the volcano all in their path is destroyed, as happened to St. Pierre in Martinique in 1902, when the volcano Mount Pelée erupted, killing 35,000 people. Tuff (hardened ash) of various grain sizes can be found on the Wrekin. Volcanic bombs occur in coarse ash on Lawrence's Hill and very fine ashes, including ignimbrites, on the Ercall and the south west ridge of the Wrekin, where some andesitic lavas and tuffs also occur.

Up to 1,500 metres of tuffs and lavas occur on the Wrekin. The fact that rhyolites are common suggests that the volcanicity would have been very explosive. Rhyolite is today found as a sticky lava which tends to block up volcanic vents and cause build ups of pressure resulting in violent explosions of the Mount St. Helens type in 1980. The rhyolite lavas around the summit of the Wrekin, some of which may be ash flows, show beautiful fine banding, known as flow banding, caused by the streaking out of minerals as the lava slowly flows. The rock itself is very fine-grained since it cooled very quickly and the crystals had little time to grow to any size.

Towards the end of this volcanic episode about 600 million years ago a number of molten masses of igneous rock were injected into the already hardened lavas and tuffs in the form of vertical thin sheets known as dykes. These on the Wrekin have a composition similar to basalt and are called dolerite, a dark-coloured igneous rock. They are up to 3 m across, occasionally thicker. When these occur in large numbers they are known as dyke swarms, and this type of feature is well known in the north west of Scotland, the Isles of Mull and Arran. They indicate a late stage of volcanic activity and are a type of intrusive (i.e. pushed in) igneous rock.

A larger intrusive body of granitic type of rock is intruded into the volcanic rocks of the Ercall. This is a fine-grained pink rock known as a granophyre. Under the microscope it shows a beautiful intergrowth of quartz and feldspar crystals on a fine scale resembling hyroglyphics and known as a graphic texture. Again, this body is a late stage intrustion dated as being 560 or possibly 533 million years-old. So the Wrekin and Ercall show classic features of volcanic and intrusive igneous activity, but it must be reiterated that no vents have been found anywhere and so the Wrekin is really an erosional remnant, now faulted and folded, of sheets of lava and tuff which would have covered a much larger area. The vents could have been anywhere in the area, so we cannot call the Wrekin an extinct volcano, even though from the Cressage area it does have a conical appearance.

In this area the Church Stretton Fault is really a system of faults with many branches, and other quite large volcanic outcrops occur between its two main branches, the Brockton Fault and the Wrekin Fault (Figs. 4 and 5), forming Wrockwardine Hill and Charlton Hill. To the south-west the band narrows and crosses the River Severn near Cound, and then the next conspicious exposure of volcanic rocks is the well known hog back of the Lawley bounded on both sides by branches of the Church Stretton Fault. The Lawley has no rhyolite lava but a large thickness of fine-grained tuffs and, at its south west end, good exposures of andestic lavas. These ancient lavas on the Lawley have small holes in them called vesicles, where gases bubbled up to the surface of the flow. Many lava flows have vesicular tops since in the volcano the molten material is under pressure, and when it comes out onto the surface it's a bit like opening a bottle of beer — all the bubbles of dissolved gases rush to the top. Sometimes these old vesicles are filled with minerals and then they are called amygdales, and can be very attractive to look at. So strictly the Lawley lavas are amygdaloidal andesites.

The line of volcanic hills now continues south west to form the eastern side of the Church Stretton valley, and next we come to Caer Caradoc, the highest of these hills at 460 m, crowned with its famous Iron Age hill fort. This beautifully shaped hill has a profile not dissimilar to that of the Wrekin, a long hog back when viewed from the west (Plate 3) and a conical

Plate 4. The Church Stretton Valley from the southern end of Caer Caradoc looking south. The valley follows the line of the main Church Stretton Fault which downfaults Silurian strata into the bottom of the valley. Ragleth Hill is on the left (Uriconian Volcanics) and the Longmynd on the right.

shape when viewed from the north-east (Plate 5). On Caer Caradoc (Fig. 9) it is possible to work out a sequence of lavas and tuffs, but lavas predominate — andesites and rhyolites, and also a new type not described before, basalt. This, the well-known rock type of the Giant's Causeway in Ireland is a very common type, particularly today in the great oceanic volcanoes of Hawaii and Iceland, and was very abundant in the past. Basalt was the fundamental rock type of the primitive Earth before the granitic continents were formed. Caer Caradoc shows a complete suite of lavas of the most well-known chemical compositions. One feature which is difficult to explain is the apparent scarcity of flow banding in the rhyolites when compared with the Wrekin. The basalt lavas are amygdaloidal in some places, and a number of dolerite dykes occur, and some larger masses of the same rock type.

The line of hills continues south west, on the east side of the Church Stretton Valley, to form Helmeth, Hazler and Ragleth Hills, but on these the rocks are poorly exposed and difficult to interpret. Ragleth Hill appears to be a great thickness of fine-grained tuffs, the Ragleth tuffs.

All these hills from the Lawley to Ragleth Hill are bounded on their north west sides by the main Church Stretton Fault, which the line of the Church Stretton valley follows. This is referred to as a normal fault, in that the rocks have simply been pushed down on one side vertically with respect to the other side. In this case the western side has been pushed down — we say the fault is downthrowing to the west (Fig.8). Many books have in the past referred to the Church Stretton valley as a rift valley, but this is caused by an area of the crust being dropped down between *two* parallel faults (Fig. 8). A graben is the technical term for this, and the Rhine graben and the East Africa Rift valley are good examples. In the Church Stretton valley there is only *one* fault, a normal fault (Figs. 11 and 19) and so the valley, which does follow the line of the fault (Plates 3 and 4), can only be called a fault valley, not a rift valley.

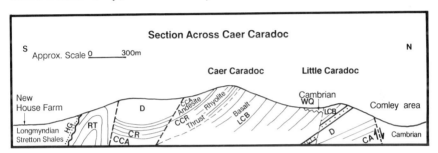

Fig. 9. North-south diagrammatic cross section through Caer Caradoc showing the various Uriconian Volcanics. Note that all three chemical suites are present as lavas — rhyolites, andesites and basalts. CA, Comley Andesites; LCT, Little Caradoc Tuffs; LCB, Little Caradoc Basalts; CCR, Caer Caradoc Rhyolites; CCA, Caer Caradoc Andesites; CR, Cwms Rhyolites; RT, Ragleth Tuffs; D. Dolerite; HG, Helmeth Grit; WQ, Wrekin Quartzite.

From Ragleth Hill the Church Stretton Fault can be traced south west towards Marshbrook and Horderley, and then at Wart Hill we come to the final exposure of Uriconian Volcanics in the county. This is an isolated conical shaped hill with a very faulted outcrop of volcanic rocks. Wart Hill has been famous in the past for its marvellous geological view of the surrounding countryside, but forestry in the 1970s somewhat obscured it, and the beautiful clump of Scots Pines on the summit was badly damaged by a severe gale on New Year's Day 1974. Further south west the Uriconian Volcanics occur again along the Church Stretton Fault at Old Radnor on the Hereford-Powys border.

The volcanic hills so far referred to are almost in a straight line north east-south west from the Wrekin to Wart Hill, all along the line of the Church Stretton Fault. However, like any really large fault, the Church Stretton Fault is not just one but a series of subparallel faults, which is often called the Church Stretton Fault system. The main branch of the fault, as already mentioned, bounds the volcanic hills on the north-west along the line of the Church Stretton valley. Further east, within 2 km or so, there are two other branches of the fault, which bound Caer Caradoc and the Lawley on their eastern sides. These branches of the fault separate Caer Caradoc and Helmeth Hill from another large area of Uriconian volcanics to the east forming Hope Bowdler, Willstone and Cardington Hills. This little known track of volcanic hill country covers quite a large area, six square kilometres (Fig. 4), and contains a variety of volcanic rocks, including rhyolite lavas which form the conspicuous crag of the Gaerstone on the south side of the Hope Bowdler Hill, above the Church Stretton to Much Wenlock road. A number of low angle faults (called thrusts, in contrast to most faults which have steep angles of inclination) separate these rocks from Ordovician rocks to the north, and to the east and south they are overlain by Cambrian and Ordovician rocks.

The Western Uriconian Outcrop

We have so far only dealt with the volcanic rocks east of Church Stretton, but another less conspicuous outcrop occurs to the west of the Longmynd along the line of the Pontesford-Linley Fault. These are often referred to as the Western Uriconian Volcanics, but as will be shown later they are simply the eastern outcrop repeated by folding. At the northern end of the Pontesford-Linley Fault near Pontesbury is the conspicuous Earl's Hill, with Pontesford Hill as its northern subsidiary summit. Again, like the Wrekin and Caer Caradoc, it has a tantalising volcanic cone-like

Plate 6. Earl's Hill from the north-east. The hill is formed of Uriconian Volcanics of the western outcrop. Upper Coal Measures in the foreground. Western Longmyndian sandstones on the left.

appearance from the north-east (Plate 6), but from the south-east it is a hog back ridge. Tuffs and lavas of rhyolitic through to basic types occur, and there are large dolerite intrusions. South west of Earl's Hill the Pontesford-Linley Fault can be traced for 18 km to Linley and along it (Fig. 4) there are sparodic outcrops of volcanics, but generally the exposures are poor, the most southerly being on Linley Hill.

Throughout this account I have given no definite thicknesses for the Uriconian Volcanics. This is because it is very difficult to work out a complete thickness on hills where the rocks are folded and faulted. Not only that, but each hill is isolated from another and, for example, detailed sequences worked out on Caer Caradoc, where 1500 m of volcanics are probably present, cannot be directly compared with, say, the Wrekin or the Lawley. The thickness over Shropshire might be expected to be in excess of 1500 m but we cannot be certain of the exact figure.

The rocks of the Longmynd

The final few million years of the Precambrian eon in Shropshire, approximately 590-575 million years ago, saw the formation of the Longmyndian Supergroup of sedimentary rocks, which today forms the large deeply dissected moorland plateau of the Longmynd (Plates 7 and 8),

Plate 7. The Longmynd from Caer Caradoc, clearly showing it to be a dissected plateau.

30   Geology in Shropshire

Plate 8. The Longmynd from the east, showing the deeply incised valleys (batches). Cardingmill Valley is in the centre.

west of Church Stretton, rising to 517 m, as well as areas directly to the west and north of the Longmynd itself (Fig. 4).

Sedimentary rocks are formed by the laying down of layers (strata) of fragmentary material, formed from the erosion of pre-existing rocks. The layers of strata, or beds, are separated by distinct lines known as bedding planes along which the rock will usually split. Throughout geological time sedimentary rocks have been laid down under water, in seas or rivers and lakes, and on land, e.g. dune sandstones, and great thicknesses can accumulate. Sandstones, shales (bedded mudstones) and limestones are well-known examples.

The Longmyndian is a truly remarkable group of rocks. They were deposited just after the close of the Uriconian Volcanic episode, and do, in fact, contain some thin volcanic bands. More particularly they represent a sequence of shallow water marine sediments laid down in a sinking trough between the Church Stretton Fault and Pontesford-Linley Fault, and thus were formed over a very limited geographical area. Eight thousand metres of these sediments accumulated and hardened, and are now found as shales, mudstones and sandstones. After the rocks were laid down (Fig. 10) they were folded into a very large downfold or syncline about 10 km across (Figs. 10 and 11). The fold was tightly compressed, and on a very large scale, so that the layering, or bedding planes, in the sedimentary rocks, originally horizontal, are now very steeply inclined, in some cases almost vertical (Plates 9 and 10). This inclination of bedding planes is

Shropshire's Oldest Rocks 31

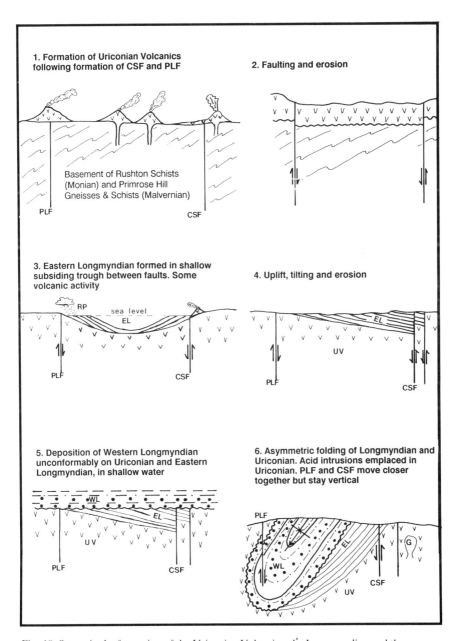

Fig. 10. Stages in the formation of the Uriconian Volcanics, the Longmyndian and the formation of the Longmynd fold structure and associated faults. PLF, Pontesford-Linley Fault; CSF, Church Stretton Fault System; UV, Uriconian Volcanics; EL, Eastern Longmyndian; WL, Western Longmyndian; G, Granophyre; RP, Rain pits. (After Toghill and Chell, 1984.)

called dip, so the Longmyndian rocks are said to dip steeply, and the direction of inclination is usually to the north-west. The steep dips are seen well in the numerous rock exposures (outcrops) in the steep-sided valleys (batches) of the eastern Longmynd. This enormous syncline is not one which you can actually see on the ground; it is mapped out by proving the repetition of sedimentary rocks on either side of a fold axis, and also the appearance of the Uriconian Volcanics on either side as well (Figs. 10 and 11). So tight was the folding that the western side, or limb, of the fold is actually overturned, and the sequence of rocks is inverted as seen on the left side of Fig. 11.

So we envisage, about 590 million years ago, a barren volcanic landscape covered with extinct, and occasionally dormant volcanoes. No life on land, and only marine algae as the abundant marine life form. On to this barren landscape a shallow sea transgressed between the two fault lines. We were about 60° south of the equator at this time. As the land on either side of this arm of the sea was eroded away, the sediments derived from the volcanic landscape were washed into the shallow sea and deposited as muds and sands. The arm of the sea slowly subsided and the sediments were turned into rocks as they were compressed by burial under more sediments — sandstones and shales were formed. Many of the fine-grained shales appear almost like tuffs because the material they are made of was derived from eroded Uriconian tuffs. This is certainly true of the earliest Longmyndian Stretton Shale Formation. The sea was to eventually receive 8000 m of sediments, probably more, because this is the present thickness of rock after compression, and suggests a much greater thickness of original sediment which had to be dried out and compressed to form solid rock. Whatever the thickness of sediment received, it did not require a sea of enormous depth. It simply subsided at the same rate at which the sediments came in. In fact it was always shallow, as can be seen by some remarkable sedimentary structures preserved in some of the shales, originally soft muds. Sometimes shallow water muds accumulated on very gently sloping shorelines, so that when the tide went out great mud flats were exposed. At certain times of the year the tide would not return to the higher parts for several days, so if the mud dried out in a strong sun, a slightly hardened crust would form. If then an April shower passed over, the large rain drops could make crater-like impressions (rain pits) in the semi-hardened mud. When the shower passed away and the sun came out, the impression would harden, and the return of the sea later could cover these with a fine mud without destroying them, in fact fossilising them for ever. One of the great thrills of the Longmyndian is to split open shales and see the obvious impressions of fossilised rain pits 590 million years-old. With an average diameter of 2-3 mm, many have obvious raised rims, while others are elliptical in outline, suggesting the rain drops came down

Shropshire's Oldest Rocks 33

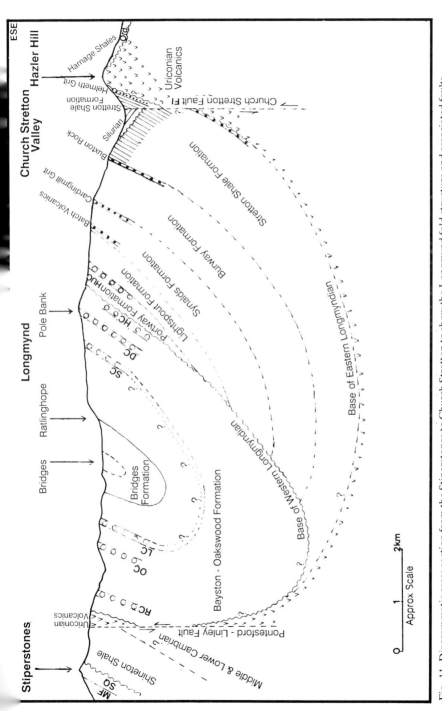

Fig. 11. Diagrammatic cross section from the Stiperstones to Church Stretton to show the Longmynd fold structure and associated faults. Thickness of beds not to scale. MF, Mytton Flags; SQ, Stiperstones Quartzite; RC, Radlith Conglomerate; 'HC', Haughmond Conglomerate, appears 7 km north of line of section; OC, Oakswood Conglomerate; LC, Lawn Hill Conglomerate; SC, Stanbatch Conglomerate; HUC, Huckster Conglomerate; DC, Darnford Conglomerate. (After Toghill and Chell, 1984.)

at an angle as they often do in squalls. Of course, if the shower was torrential all the rain pits would coalesce, and so we only find good examples where the mud flats have just been on the edge of a shower. These can be produced today both in the laboratory and in modern estuarine mud flats. If you are lucky you could go to one of these areas and see it happening. The rain pits are most common in the Synalds Formation of the Longmyndian, which is made up of purple and greyish purple shales, and this formation also includes numerous beds full of types of ripple-marked surfaces indicative of a beach environment. Indeed, many of the ripple-marked shales are covered in rain pits as well. So the sea was a shallow one, but in receiving all the material that it did, it produced a group of sedimentary rocks unique to southern Britain.

The full sequence of the Longmyndian is shown in Table 2.

The succession of sandstones and shales is divided into two major groups, the Wentnor Group, or Western Longmyndian, separated from the Stretton Group, or Eastern Longmyndian, by an unconformity. The Western and Eastern Longmyndian used to be called the Red and Grey Longmyndian owing to the dominant colours of the two major rock groups. There are other marked differences between these two groups. The term unconformity is familiar to all geologists but difficult to explain to the amateur. In an ideal situation during the laying down of sedimentary rocks every moment of time would include some amount of deposition. In fact this is rarely true, and there are many periods of time on the sea floor when no material is being deposited, and in many cases it is actually being washed away by currents as fast as it is laid down. A bedding plane in a

Table 2. The Longmyndian Sequence.

Wentnor Group (Western Longmyndian)	
Bridges Formation – purple sandstones	600-1,300 m
Bayston-Oakswood Formation – purple sandstone with conglomerates	1,200-2,400 m
Unconformity	
Stretton Group (Eastern Longmyndian)	
Portway Formation – green and purple sandstones, Huckster Conglomerate at base	180-1,700 m
Lightspout Formation – grey green sandstones and silstones	520-820 m
Synalds Formation – purple shales and sandstones with the Batch volcanics (tuffs) at top	480-850 m
Burway Formation – greenish grey sandstones and shales, with the Cardingmill Grit at top and Buxton Rock (tuff) at base	600 m
Stretton Shale Formation – greenish grey shales with Helmeth Grit at base	?-960 m

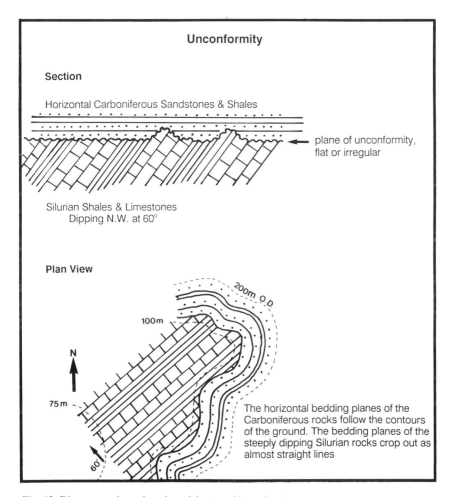

Fig. 12. Diagrammatic explanation of the term Unconformity.

normal sequence of sandstones indicates a brief pause in deposition. However, if the break in deposition is, in terms of time, thousands or millions of years, or represented by earth movements, then this means something is missing from the geological record as represented by sedimentary rocks. *Time* is unrecorded by the rocks. This is one meaning of an unconformity.

As a *physical* feature, an unconformity is often shown by a discordant relationship between two sets of beds (Fig. 12), which dip in different directions. The unconformity is in this case a plane or irregular surface representing a time of uplift and erosion of the pre-existing beds before the next set of beds was laid down. An unconformity can be shown to be an

Plate 10. Steeply dipping Longmyndian rocks in Cardingmill Valley.

erosion surface, an old shoreline for example. An angular unconformity of this type often separates rocks of one geological period from another, indicating that, for instance, at the end of the Ordovician period in Shropshire there was a period of uplift and erosion prior to the deposition of the Silurian rocks. The late Ordovician is not represented by rocks, the time is represented by folding and erosion. When an unconformity is seen on a geological map, a particular set of rocks is seen to cut across the sequence of older rocks beneath the unconformity. Have a look at the Geological Survey Sheet 166 (Church Stretton) at the base of Silurian (*Pentamerus* Beds) between Acton Scott north west to Plaish, and you can see the base of the Silurian rests on various rock formations of Ordovician and Cambrian age and even touches the Precambrian near Gilberries.

Thus, within the Longmyndian is a major unconformity separating the Wentnor Group (Western Longmyndian) from the underlying Stretton Group (Eastern Longmyndian). During the time represented by the unconformity the Stretton Group was folded and eroded (Fig. 10). The Wentnor Group was then deposited so that it rests on various formations with the Stretton Group (Figs. 10 and 11), and, in fact, in places rests on the Uriconian Volcanics, so that the whole 4,300 m of the Stretton Group has been eroded away in, perhaps, a period of four million years, or a rate of 1 mm per year. A rate of erosion of 1 cm per year would lessen the time interval represented by the unconformity to 400,000 years. Choosing the rate of erosion which you think is appropriate is very difficult, and is dependent on climate, rock types, etc., but at least we can see that the time intervals involved can vary enormously. We shall see many other examples of unconformities in the rest of geological time in Shropshire.

The base of the Longmyndian, the Stretton Group, is only seen east of the Church Stretton Fault and is marked by a coarse sandstone, the Helmeth Grit, 30 m thick, overlying the Uriconian Volcanics, probably unconformably. The Helmeth Grit is very ashy (tuffaceous) and could represent a dying phase of Uriconian Volcanicity. It is the basal member of the Stretton Shale Formation which is predominantly greenish grey shales up to 960 m thick and exposed on either side of the Church Stretton Fault. On the east side of the fault, the Stretton Shales contain a large number of thin, pale-coloured clay bands up to 5 cm thick. These are bentonites, clays derived from the weathering of fine-grained ashes, and are exceedingly useful for radiometric dating purposes. These soft clays are in contrast to the more normal, very hard volcanic tuff horizons, a number of which, up to 4 m thick, occur higher up in the Stretton Group, indicating other dying phases of Uriconian Volcanism. One of these, the Buxton Rock, at the base of the Burway Formation, is exposed in a small quarry behind the Yew Tree Inn in All Stretton. It is a very fine-grained, almost glassy, grey siliceous tuff, probably an ignimbrite, the product of a glowing gas cloud eruption. The tuff itself is shot through with quartz veins. Further west, and higher in the sequence, the Batch Volcanics contain a number of thin rhyolitic and andesitic tuffs, and these are just above the level in the Synalds Formation of purple shales with abundant ripple marks and rain pits. Some rain pits appear to occur in the lowest tuff band, the so-called White Ash. There is no reason why this ash could not have been deposited as a fine mud on the edge of the shoreline which would allow rain pits to form.

The rest of the Stretton Group comprises sandstones, including the well-known Cardingmill Grit, 36 m thick in Cardingmill Valley, and shales variously coloured grey, green and purple. Many of the sandstones are badly sorted according to grain size and were probably deposited quite quickly without time for the heavier grains and particles to be separated from the finer material. This type of sandstone is called a greywacke and is usually associated with deep water turbidity currents off the continental shelf. However there is no reason why greywackes cannot form in shallow water, if the material is rapidly deposited.

Another feature of the Stretton Group, particularly in the Burway and Synalds Formations, is graded bedding where sediment of various grain sizes settles out slowly so that the coarser material is found at the bottom and the sequence becomes finer upwards. This may occur in a bed of sandstone 6 cm thick, followed by another bed repeating the whole process, and so on, several times, indicating perhaps regular cycles of deposition. Graded bedding is perhaps the opposite to a greywacke environment.

The Stretton Group occupies most of the eastern side of the Longmynd, deeply dissected by streams flowing south east, and thus exposure of the various rock types is very good, and also all the southern part of the Longmynd. A large, mainly drift-covered, area occurs east of the main Church Stretton Fault on the flanks of the older volcanic hills in the Church Stretton Valley, and this is the only place where the base of the sequence is exposed, with the Stretton Shales, including the Helmeth Grit at their base, resting on Uriconian Volcanics. This eastern outcrop of Stretton Shales looks rather different to that of the main outcrop on the Longmynd itself, west of the Church Stretton Fault. It is not as deformed as the main outcrop and contains numerous bentonites, which are completely absent west of the Fault. Owing to the two outcrops being separated by the Church Stretton Fault, the true thickness of the Stretton Shales can only be estimated at around 960 m. It is possible that a large amount of tear faulting has occurred along the Church Stretton Fault, and the Longmyndian of the main outcrop may be part of a different geological terrane (see later) brought into the area by horizontal movement along the Fault. Faulted outcrops of the Stretton Group occur between Marshbrook and Wart Hill. North east of the Longmynd, a small patch of Longmyndian rocks at Pitchford overlain by Coal Measures, and in places stained with pitch leached from above, is part of the Stretton Group, and the Group also occurs on Lyth Hill, and Haughmond Hill north of the River Severn.

The Wentnor Group is much coarser than the Stretton Group and mainly consists of medium to coarse-grained purple, red, and occasionally grey, sandstones, as well as even coarser conglomerates (consolidated gravels). The latter have local names, but it is possible to compare the three bands in the east of the outcrop with those in the west and prove the synclinal nature of the Longmynd fold structure (Fig. 11).

The Wentnor Group forms the highest ground of the Longmynd itself, where exposure is rather poor, and also quite a large area further west around Ratlinghope, Wentnor and Pulverbatch, where the conglomerate bands form conspicuous hills. Also, in this western area the Longmyndian is cut by mineral veins of late Devonian age rich in barytes (heavy spar), together with a little copper. This important feature will be discussed more when we deal with the rocks of the Ordovician period and the extensive mineral deposits west of the Stiperstones, which are of the same age as those in the Longmyndian.

The western limit of Wentnor Group is either the Pontesford-Linley Fault, where the Longmyndian is faulted against late Cambrian Shineton Shales, or Precambrian Uriconian Volcanics, as at Earl's Hill. Further south west along the fault line near Linley there is a normal contact with the Uriconian Volcanics, and the Wentnor Group is seen to overlie unconformably the volcanics, but with an inverted contact (Fig. 11).

The Wentnor Group forms an outcrop which continues north of Pulverbatch to form the conspicuous ridge of Lyth Hill, where the conglomerate bands are up to 100 m thick, and the dip is very steep, almost vertical in places. At Sharpstones Hill, the north end of Lyth Hill, very hard greywackes between two conglomerates are quarried for roadstone, said to be the best in Shropshire. Here, the dip is vertical and the greywacke is being quarried out by following the beds downwards as far as is physically possible, as well as laterally between the two conglomerates.

Northwards the Longmyndian rocks disappear under younger Coal Measures just east of Shrewsbury, but north of the River Severn the Wentnor Group, and part of the highest Stretton Group, reappears to form Haughmond Hill, a Longmyndian inlier, which is completely surrounded by younger Coal Measures. On Haughmond Hill the exact same seam of greywacke as that of Sharpstones Hill has been quarried for roadstone.

Further north the Longmyndian rocks disappear under the Triassic North Shropshire plain, never to be seen again on the surface. We do not know how far north they could be present underground. The southward continuation of Longmyndian rocks south of the Longmynd is only seen along and within the Church Stretton Fault system in which faulted and folded outcrops of both Stretton and Wentnor Groups occur, as far south as Hopesay Hill west of Craven Arms. Small outcrops occur further south near to the fault line at Pedwardine in north Herefordshire, at Old Radnor associated with Uriconian Volcanics, and also possibly at Huntley Quarry in Gloucestershire. The Wentnor Group does appear to have covered a slightly larger area than the Stretton Group. There is a very critical exposure in the Cwms area on the south flank of Caer Caradoc where Cambrian rocks are supposed to overlie purple sandstones of the Wentnor Group. If this is true, then this is the only place where the age of the Longmyndian can be proved to be directly Precambrian. In all other areas the Longmyndian rocks can only be directly shown to be pre-Silurian, or are faulted against Cambrian or Ordovician rocks. There are a number of places on the southern Longmynd where Silurian rocks rest with a marked unconformity on folded Longmyndian (see later). Around Hope Bowdler the Uriconian Volcanics are definitely pre-Caradoc (late Ordovician) in age, and as they are followed directly by the Longmyndian, one can also argue that the Longmyndian is pre-Caradoc. However because the Longmyndian rocks are totally different to anything in the Cambrian and Ordovician, they are usually accepted as being part of Precambrian age, and were folded to the present structure at the close of that era. There is a possibility that the Longmyndian, west of the Church Stretton Fault, could be part of an exotic early Cambrian terrane brought into the area by long distance tear faulting along the Church Stretton Fault and Pontesford-Linley Fault, and folded in late Ordovician times along with the Shelve area. Terranes feature a good deal in present day concepts of plate tectonics. There are marked differences between the Stretton Shale Formation either side of the Church Stretton Fault as mentioned earlier. The terrane idea can still be applied, even if the balance of evidence suggests that the Longmyndian is Precambrian.

The Longmyndian fold structure

Figures 10 and 11 show the suggested mode of formation of the Longmyndian sediments, and the way in which they have been folded into a tight, overturned syncline. This folding is probably of late Precambrian age (but see above), causing the whole of Shropshire to be raised above sea level at this time. The fold structure lies between the Church Stretton Fault and the Pontesford-Linley Fault and includes the Uriconian Volcanics as part of the syncline, thus explaining how they appear on both sides of the syncline, on either side of the Longmyndian outcrop. The core of the syncline contains the youngest formation, the Bridges Formation, and either side (Fig. 11) are repeated conglomerate bands. The western limb of the fold is inverted, so that at places near Linley one can see the inverted contact of the Uriconan Volcanics apparently resting on the Wentnor Group (Fig. 11). This contact led earlier workers to suggest there were two volcanic formations, one below and one above the Longmyndian. Now we accept that inversion of strata can occur in tight folds, one must always be careful in studying areas of tight folding and steep dips to work out exactly which way up the strata are, the youngest rocks are not necessarily at the top of the sequence!

A number of faults which cut the Longmyndian are the same age as the folding. The most conspicuous is the Longmynd Scarp Fault (Plate 11) which forms the western boundary of the Longmynd hill mass (not the Longmyndian outcrop) at its southern end for 8 km. This very steep scarp is 200 m high and forms an excellent site for the Gliding Club on top of the Longmynd. This fault also cuts the Silurian rocks at the foot of the scarp and was probably reactivated during the Tertiary period, having probably been initiated as a Precambrian fault. A number of dolerite dykes cut the Longmyndian and the fold structure, and are of very late Precambrian age.

People have searched for over a hundred years for fossils from the Longmyndian. Salter, in the late 1800s, originally described the rain pits as worm tubes, and named them *Arenicolites*. The famous Precambrian fossil *Charnia* from Charnwood Forest is a sea pen, 12 cm long, from a shale sequence within volcanics of Uriconian type. No such macro fossils have

Plate 11. Aerial view of the spectacular Longmynd Scarp Fault which forms the western margin of the southern Longmynd. Wenlock Edge and the Clee Hills are in the distance
(Photograph taken by Chris Musson.)

been found in the Longmyndian, but perhaps shale sequences in the Uriconian Volcanics, which do occur, may yield fossils. In the 1980s the Longmyndian yielded micro fossils. Fine filamentous algal mats were found in the Stretton Shale and Lightspout formations, and occurrences of the enigmatic colonial animal *Arumberia* have been reported, the latter leaving impressions often looking like sedimentary structures. Many quite strange small ripple marks in the purple shales of the Synalds Formation look just like fossil plants. Medusoid fossil impressions may occur, and very fine sinuous ridges have been interpreted as the linings of worm tubes.

The age of the Shropshire Precambrian rocks

The absolute age of the Shropshire late Precambrian rocks is a continual matter for discussion amongst geologists. The problem is that for the last thirty years we have taken the base of the Cambrian period at 570 million years ago, and so by definition the Uriconian Volcanics and the Ercall Granophyre (Figs. 15 and 16), which are quite demonstrably 'pre' the Cambrian Wrekin quartzite, must be of Precambrian age, and thus older than 570 million years. However in recent years a radiometric date from the Uriconian Volcanics of 558 ± 16 Myrs (million years) has been obtained and 560 and 533 ± 13 Myrs from the Ercall Granophyre. These are all 'Cambrian' dates, which poses a problem. Either these rocks are Precambrian and the base of the Cambrian will have to be placed at around 530 million years ago, or the rocks are early Cambrian.

The Longmyndian sediments have also yielded absolute ages from fission track dating of 530 Myrs which is well into the Cambrian, and they are, of course, quite definitely younger than the Uriconian Volcanics, and the youngest of the Shropshire Precambrian sequence. Bath (1974) determined a maximum age of deposition for the Longmyndian of 590 Myrs based on Strontium initial ratios in shales. Of course, it may be that all these dates are incorrect, i.e. the laboratory tests contain errors, or it could be that the dates obtained are not the ages of the rocks, but later 'metamorphic' events which have affected them, e.g. slight heating in the Cambrian. A date of 536 ± 8 Myrs biotite cooling age for the Rushton Schist (Patchet *et al.*, 1980) is certainly not the age of the Rushton Schist itself, since it is a garnet grade rock and could not have been metamorphised to this grade at this time without other surrounding rocks being affected. This date possibly represents again a Cambrian event.

Until all these problems are resolved it is best to accept the conventional date for the base of the Cambrian at 570 Myrs, and thus in this account I have used a date of 700 million years for the Rushton Schist, 650-600 million years for the Uriconian Volcanics, and 590-575 million years for the Longmyndian. This allows five million years for the late Precambrian folding of the Longmyndian and Uriconian into the famous synclinal structure and for its uplift and erosion.

We have now finished our look at Shropshire's oldest rocks which provide a fascinating insight into conditions on a primitive Earth 700-570 million years ago.

Chapter 4
The first signs of abundant Life – the Cambrian period in Shropshire

The Cambrian System was first named by Adam Sedgwick in 1835 to cover the ancient rocks he had been studying in North Wales (Cambria), which contained very primitive forms of life, particularly the strange creatures looking somewhat like woodlice that we call trilobites. The Cambrian period is the time interval in which the rocks of the Cambrian System were laid down. We have discussed earlier the difference between the term period and system. The Cambrian is the first period in which we find abundant forms of life fossilised (Fig. 17), but this is probably only a reflection of the new-found ability at this time for animals to secrete hard shells capable of being fossilised, from increased calcium carbonate levels in sea water. The rocks prior to the Cambrian belong to an era which was formerly called Azoic (without life), but subsequent to the discovery of evidence of life in these older rocks some are now called Proterozoic (former life), or for want of a better word, Precambrian — i.e. the whole time interval prior to the Cambrian. We usually estimate that the period started about 570 million years ago, and finished about 490 million years ago. It was a time when the shallow continental shelf seas contained a great variety of invertebrate life forms, as well as marine plants, but the land was devoid of life completely.

We can subdivide the Cambrian period into four smaller time divisions which are, in ascending order: the Comley Series, St. David's Series, Merioneth Series and Tremadoc Series. The first three of these new terms equate to the old divisions of Lower, Middle and Upper Cambrian. The

Comley Series is named after Comley Quarry near Church Stretton, where the first British Lower Cambrian trilobites were found in 1888. The position of the Tremadoc Series has always been in dispute. In Britain many geologists, along with the rest of the world, now place the Tremadoc within the Ordovician. However the British Geological Survey still place it at the top of the Cambrian, as defined by Lapworth when erecting the Ordovician System in 1879. In this book I follow the traditional view.

The modern ideas of plate tectonics now applied to the older rocks have led to the palaeogeography of Cambrian times being explained in terms of the presence of an ancient ocean in which the British Cambrian rocks were formed in widely separated parts of the world, some near the equator and some near to the Antarctic Circle. This, the Iapetus Ocean, will now be discussed in more detail.

The Iapetus Ocean

The name Iapetus (or Proto Atlantic) Ocean was first used in the 1960s for a wide oceanic area which was thought to separate northern and southern Britain, and other areas around the present Atlantic, during the Cambrian, Ordovician and Silurian periods. Even in the early parts of this century, it was noticed that the Cambrian fossils of north west Scotland were slightly different from those of Wales and Shropshire. Although the trilobite *Olenellus* occurs in both areas, there are slight differences in the species — the same sort of differences between, say, the Indian and the African elephant today. These modern differences were discussed in great detail by Darwin, who suggested the idea of faunal provinces — closely related animals evolving in isolation, and thus appearing different from each other.

So in the Cambrian trilobites evolving on either side of a wide ocean would have slight differences. They would not be able to cross the ocean to interbreed as they were shallow water creatures, and so two separate communities would evolve. We now recognise Cambrian faunal provinces and Fig. 13 shows which parts of the British area were on opposite sides of the ocean during the Cambrian, and you will also notice that one half of Newfoundland was on the north side and one half on the southern side.

During the Cambrian the actual name Iapetus Ocean can only really be used for the area between Scotland, Greenland and parts of North America (Laurentia), and Scandinavia (Baltica) (Fig. 13). The ocean was probably born when Baltica separated away from Laurentia during the early Cambrian, with an ocean ridge forming between the two. Southern Britain and east Newfoundland were already in the southern hemisphere at 60°-70° S, attached to the Gondwana supercontinent of Africa, India, South America, etc. (Fig. 13). Thus the Shropshire area close to the

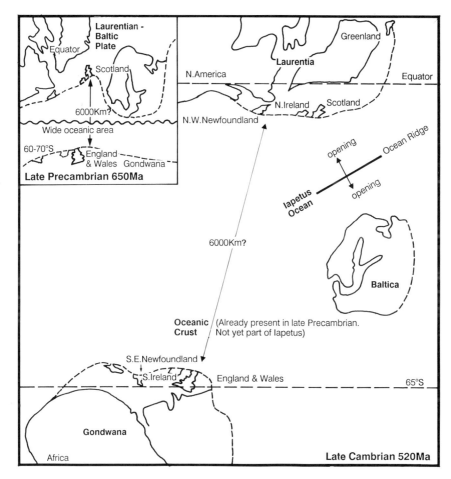

Fig. 13. Cambrian paleogeography and the formation of the Iapetus Ocean (Proto-Atlantic) as Baltica splits away from Laurentia. Southern Britain and south east Newfoundland are attached to Gondwana 6000 km away, and the southern oceanic area in between cannot at this time be called Iapetus. Inset shows the situation in the late Precambrian, with Laurentia and Baltica joined together before the formation of the Iapetus Ocean.

Antarctic Circle was probably 6000 km away from Scotland, which lay near to the equator. We cannot use the term Iapetus for this wide oceanic area in the Cambrian.

During the Cambrian the Iapetus Ocean widened between Laurentia and Baltica, but then in the early Ordovician, or even late Cambrian, it started to close, with subduction zones on either side (Figs. 20 and 21). At the same time southern Britain and east Newfoundland broke away as a microcontinent from Gondwana and started to move northwards with a subduction zone on its northern margin, and a new ocean, the Rheic

Ocean, forming between it and Gondwana. All these three subduction zones caused widespread volcanic activity, mainly of the island arc type, during the Ordovician, and it is in the Ordovician that we can now apply the term Iapetus Ocean to the whole area between Laurentia, Baltica, and the southern Britain-east Newfoundland microcontinent, with the name Tornquists's Sea given to the oceanic area between southern Britain and Baltica (Fig. 20).

During the Ordovician not only can one recognise faunal provinces represented by different types of graptolites and trilobites, but also different types of sediments which give an idea of the latitudes. Thus the Lower Ordovician of north west Scotland (the Durness Limestone) is a tropical carbonate formed near the equator whereas the equivalent rocks in southern Britain appear to be temperate sandstones (such as the Stiperstones Quartzite) formed around 60°S, and the Iapetus Ocean in the early Ordovician (Arenig) would thus be about 5,000 km wide (Cocks and Fortey, 1982) (Fig. 20).

McKerrow and Cocks (1976) devised a method of calculating the changing width of the Iapetus Ocean by studying progressive animal migration across Iapetus. They suggested certain groups of animals would be capable of crossing the ocean before others, planktonic graptolites first, and trilobites and brachiopods next, etc. They studied larval stages of modern brachiopods with a lifespan of seven to fourteen weeks which could thus possibly cross an ocean when it was between 2,000-4,000 km wide. They also worked on subduction rates suggesting that a rate of subduction of 4 cm per year would have subducted 7,600 km of ocean crust between the late Cambrian and mid-Silurian. The evidence from pelagic larval migration and subduction rates both suggest that the ocean had a minimum width of 2,000-3,000 km by the end of the Ordovician.

The ocean continued to close during the later Ordovician, and McKerrow and Cocks (1986) refer to the southern micro-continent of east Newfoundland and southern Britain as Avalonia (Fig. 20), named after the Avalon peninsula in Newfoundland. A new ocean, the Rheic ocean, is now present between Avalonia and the southern continents of Africa, South America, etc. (Gondwanaland of Wegener).

By the end of the Ordovician, Tornquist's Sea has closed and the collision of Avalonia and Baltica caused the late Ordovician orogeny over southern Britain often called the Taconian. However the term Taconian is really best applied to events in North America, and as we have widespread earth movements in Shropshire of late Ordovician (Ashgill) age, as shown by the folding of the Shelve area, I shall call this orogeny the Shelveian. It is one of the most important tectonic events in southern Britain.

After the collision of Avalonia and Baltica they continue to move northwards as one unit during the Silurian, towards Laurentia which is stationary near to the equator (Fig. 33). The northward movement is caused by the spreading Rheic Ocean to the south (Fig. 33). The Iapetus Ocean finally closes at the end of the Silurian and Avalonia/Baltica collides with Laurentia to cause the main Caledonian orogeny which affects Scotland and Northern Ireland and many parts of Scandinavia and North America in the late Silurian. The two halves of the British Isles are now welded together (sutured) along a line near to the Solway Firth area, and the two halves of Newfoundland join up, and Scandinavia collides with Greenland and North America. A vast mountain chain, the Caledonian mountains, had now been formed by the early Devonian over north west Britain and other areas (Fig. 39). No late Silurian earth movements affect southern Britain and Shropshire, and the term Acadian might be more appropriate than Caledonian for the main early Devonian orogeny in Wales and the Lake District, and minor earth movements in Shropshire (see later).

The spreading Rheic Ocean to the south (Figs. 33 and 39) was itself to play an important part in the formation of the folded and metamorphic rocks of south west England and south Wales, as it opened and then closed to cause the Variscan orogeny in late Carboniferous times, as will be explained later.

Three main lines of evidence therefore support the presence and the closure of the Iapetus Ocean: the volcanic rocks proving subduction zones, and therefore oceanic trenches and island arcs; the faunal provinces as shown by the differing trilobite and graptolite faunas on either side in Cambrian and Ordovician times; the climatic evidence of the sedimentary rocks which show contemporary formation in tropical and temperate latitudes. The ocean was becoming narrower in Silurian times so that the earlier faunal provinces ceased to exist as animals were able to cross the sea. Further aspects of the evolution of the Iapetus Ocean will be discussed when describing the Ordovician and Silurian periods.

As explained above, and during an earlier summary of the theory of plate tectonics (Fig. 1), the evidence for subduction zones having been present in the past is the presence in the older rocks of igneous and particularly volcanic rocks, metamorphic rocks and major faults, suggesting past earthquakes activity. The identification of faults associated with subduction zones is perhaps the most difficult of these to identify. Since no volcanic or metamorphic rocks of Cambrian age occur in southern Britain we can assume that no subduction took place here.

Over southern Britian a period of late Precambrian earth movements was followed by the spreading of a Cambrian sea over the land, a phenomenon called a marine transgression. Sea level did not rise, the continental margin subsided. Everywhere there is a marked unconformity at the base of the Cambrian indicating that the Precambrian rocks were

Diachronism, Overlap and Overstep.

Sea transgressing from left to right.

1 Diachronism

As the sea is transgressing from left to right the conditions for the formation of bed A move to the right, and at the same time bed A is becoming younger. At Point Z, A is 1 million years younger than it is at V. Bed A is a diachronous bed, and so are beds B - E.

2 Overlap

Each successive bed spreads over a wider area than the previous one. The situation during a marine transgression.

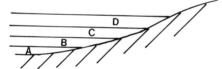

3 Offlap

Each successive bed is spread over a smaller area than the previous one. The situation during a marine regression.

4 Overlap with Overstep

A common situation along an unconformity. The series A - D shows successive overlap and at the same time comes to rest on successively older beds of the series W - Z. e.g. bed B oversteps bed Y to rest on bed X.

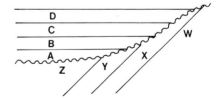

Fig. 14. Diachronism — overlap, offlap, and overlap with overstep.

uplifted and eroded before the Cambrian sea spread from the north-west, initially shallow and then deeper. The older Precambrian rocks — remember, no other rock types were present to form what we can call the southern British area at that time — were eroded away to provide the material for Cambrian sedimentary rocks to be laid down in the advancing seas. One important feature of a transgressing sea noticeable in the rocks today is called diachronism (Fig. 14). As the shoreline moved away south-eastwards in front of the advancing sea, the basal sediments would be laid down in north Wales earlier (maybe one million years earlier) than they would be in Shropshire. This means that the actual age of the base of the Cambrian rock sequence in Shropshire is slightly younger than in north Wales, and older than those in, say, the Nuneaton area of the east Midlands. Diachronism is an important feature of advancing seas, particularly in dealing with rock types dependent on certain sea bottom conditions such as reef limestones. The Silurian Wenlock Limestone contains a reef development which is clearly younger in the south than in the north, thus showing that the reef belt migrated from north to south following the part of the sea ideal for coral reef growth. In the ideal situation of a continuously transgressing sea, different belts of offshore sediments, starting with shallow water types and going through to deeper water types, will arrive one after another at a particular place as the shoreline moves.

Over Shropshire, the advancing Cambrian seas initially laid down a shallow water sandstone rich in quartz, the Wrekin Quartzite. This is hard white sandstone (not a metamorphic quartzite), and at the base is a conspicuous conglomerate, a beach gravel, now hardened, full of rounded pebbles of volcanic rocks derived from the underlying Precambrian rocks forming the coastline. As the sea deepened, a sandstone sequence followed, the Comley Sandstone, containing a very important thin limestone band with the first fossils, trilobites and types of sea shells known as brachiopods. The sea then became slightly deeper so that finer grained muds were laid down to form the well bedded Shineton Shales. Slight earth movements occurred throughout the Cambrian in Shropshire to disturb frequently, but only slightly, the pattern of continuous sedimentation, there being a number of minor unconformities and non-sequences. Movements at the end of the period caused the sea to move back westwards to a line near the Pontesford-Linley Fault. The complete sequence of Cambrian rocks in Shropshire is as shown in Table 3.

The whole sequence is very thin when compared to rocks of the same age in North Wales, where 3,500 m of continental slope deposits prior to the deposition of the Tremadoc Series compares with only 400 m in Shropshire. Conversely the Tremadoc Slates of North Wales are only 300 metres thick, a third of the thickness of the equivalent Shineton Shales in Shropshire.

Details of Cambrian rocks in Shropshire

The outcrops of Cambrian rocks broadly follow those of the Uriconian Volcanics, overlying them along the line of the Church Stretton Fault system and faulted against the whole length of the Pontesford-Linley Fault. At Lilleshall a small Cambrian outcrop is very faulted but contains Lower Comley Sandstone, and the Dolgelly Beds of the Upper Cambrian. Further south west along the line of the Wrekin Fault, Cambrian rocks are

Table 3. Cambrian sequence in Shropshire.

Tremadoc Series (Upper Cambrian)	Shineton Shales	900 m
non sequence		
Merioneth Series (Upper Cambrian) 'Dolgelly Beds'	Black Shales	5 m
	Grey (Orusia) Shales	20 m
unconformity		
St. David's Series (Middle Cambrian)	Upper Comley Sandstone	180 m
unconformity		
Comley Series (Lower Cambrian)	Lower Comley Limestones	1.8 m
	Lower Comley Sandstone	150 m
	Wrekin Quartzite	50 m

well exposed on the south east flanks of the Ercall and the Wrekin. On the Ercall, the extensive quarries in the Wrekin Quartzite show a marked unconformity at the base of the sequence where steeply dipping quartzites with a conspicuous basal conglomerate (Plate 12 and Fig. 15) rest unconformably on the Ercall granophyre or Uriconian Volcanics. The quartzite shows ripple marks indicative of shallow water, and the conglomerates contain pebbles of locally derived rhyolites, tuffs and granophyre. The pebbles are often altered to a green colour caused by the mineral chlorite which is derived from altered iron-bearing minerals, and a bright yellow colour in these basal beds is probably limonite (iron oxide). Here on the Ercall we really can see the Cambrian shoreline and imagine the sea wearing away the volcanic landscape and depositing the shallow water sandstones and pebble beds, now quartzite and conglomerate. Fifty metres of quartzite (Plate 13) are found in this area, but with no fossils apart from a few trace fossils, in this case worm borings. Near to Rushton, the quartzite rests on the Rushton Schists, and schists and gneisses at Primrose Hill.

The Wrekin Quartzite is a sedimentary rock. The term quartzite should strictly only be used nowadays for a metamorphosed sandstone with recrystallised quartz. The Wrekin Quartzite is a quartz-rich sandstone, more correctly a quartz arenite, and has not been metamorphosed. However the name is an historic one which will not be quickly changed.

Above the Wrekin Quartzite is a marked change to a greenish brown sandstone, the Lower Comley Sandstone, and this junction is well exposed in the Ercall quarries (Plate 14). The green colour is caused by the mineral glauconite, which is entirely restricted to marine environments, and is also well known from the Greensand of south east England. The Lower Comley Limestones and Upper Comley Sandstone are absent here due to faulting, and a traverse south east from the Ercall (Fig. 15) brings us on to the late Cambrian Shineton Shales well exposed on Maddocks Hill. They are named after the village of Sheinton near Cressage, which was spelt Shineton on Victorian maps. (Similar discrepancies occur in the spelling of Dolgellau — 'Dolgelly' — and Ruyton — 'Ryton'.) At Maddocks Hill the dip of the green-brown shales is almost vertical and they have been intruded by a late Ordovician igenous rock called camptonite, rich in pink feldspar, and green pyroxene and hornblende. In a simple classification of igneous rocks, this would be termed a microdiorite. The intrusion,

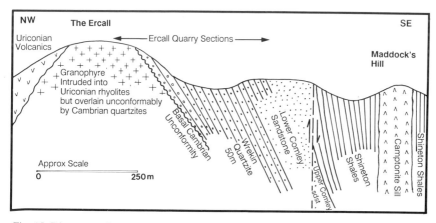

Fig. 15. Diagrammatic cross section from the Ercall to Maddock's Hill showing the basal Cambrian unconformity. Note the camptonite is a sill, even though it is vertical, as it is parallel to the bedding of the Shineton Shales.

although vertical, is parallel to the bedding and thus is a sill, in contrast to a dyke, which cuts across the bedding. The sill is about 80 m thick but can only be traced 800 m in a north east-south west direction. It has been quarried extensively for roadstone in the past, and owes its hardness to an ophitic texture, whereby the large crystals of pyroxene enclose smaller feldspars, giving the rock great strength. The intrusion has slightly baked the shales for up to 20 m on either side of the contact where they appear grey due to thermal metamorphism. The Shineton Shales at Maddocks Hill contain trilobites and, more commonly, the many branched dendroid graptolite called *Dictyonema* (Fig. 17). This colonial animal was attached to the sea

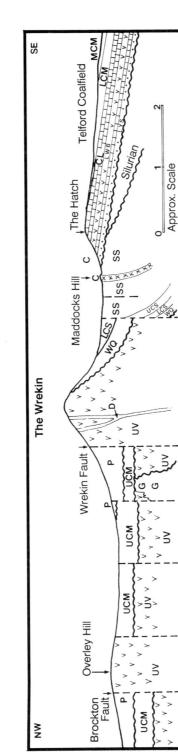

Fig. 16. Diagrammatic cross section across the Wrekin area. Thickness of beds not to scale. Permian (Bridgnorth/Lower Mottled Sandstone); UCM, Upper Coal Measures; UV, Uriconian Volcanics; G, Granite, proved in IGS Wrekin Buildings Borehold (Telford 1:25000 Sheet); D, Dolerite; WQ, Wrekin Quartzite; LCS, Lower Comley Sandstone; UCS, Upper Comley Sandstone; SS, Shineton Shales; C. Camptonite; LYS, Lydebrook Sandstone; LWB, Little Wenlock Basalt; CL, Carboniferous Limestone; LCM, Lower Coal Measures; MCM, Middle Coal Measures. (After Toghill and Chell, 1984.)

floor, and sessile dendroids gave rise to the larger group of planktonic graptolites which were abundant in Ordovician and Silurian times.

West of the Wrekin Fault, the Upper Comley Sandstone occurs but is poorly exposed. The Lower Comley Limestones have been mapped by digging trenches around Charlton Hill. The Shineton Shales outcrop widens out south of the Wrekin to cover a large area centred around Cressage and including the village of Sheinton, where Sheinton Brook provides classic exposures with late Cambrian trilobites (Fig. 17).

Further south west along the Church Stretton Fault the south east side of the Lawley (Fig. 18) is shown on Geological Survey maps as displaying a 'complete' Cambrian Sequence, but exposures are very poor. Further south west the large area of Cambrian rocks on the east side of Caer Caradoc (Fig. 19), the Comley area, is very faulted and poorly exposed, but it is here that Cobbold in the 1920s carried out a very detailed survey of the Cambrian rocks, often using trenches, and was able to find and produce a detailed sequence, and proved an unconformity at the base of the Upper Comley Sandstone, and also showed that a number of non-sequences occurred in the Upper Cambrian. Non-sequences are breaks in the deposition, but with no angular unconformities and with little erosion of the older rocks. In this case, time is not represented by rock sequences.

The Comley area is best known for the Lower Cambrian fossils which have been found in the Lower Comley Limestones at Comley Quarry, and Middle Cambrian faunas from the overlying rocks. In 1888 Lapworth described the trilobite *Olenellus* (now *Callavia*) *callavei* from the Lower Comley Limestones at Comley Quarry. This fossil was among others found there which were the first Lower Cambrian fossils described from Britain, and since then quite a large fauna of trilobites, brachiopods and gastropods has been described (Fig. 17). The Limestones are only 1.8 m thick, and the fossils are really the oldest British fossils which can be compared with other Cambrian rocks throughout the world, and thus allow international correlation and comparison of rock sequences.

The Lower Comley Limestones, although only 1.8 m thick contain five thin separate limestones, each with a distinct fauna, the lowest of which contains *Olenellus* and rests directly on the green Lower Comley Sandstone.

The fossils have been collected over many years of painstaking work and are not abundant. The casual visitor will probably find nothing. If you go to Comley Quarry, which is now a geological reserve owned by the Shropshire Trust for Nature Conservation, treat the quarry with respect and do not hammer the rock faces, as the amount of fossiliferous material is not infinite from such a thin horizon. The importance of the site is really its place in the development of the understanding of geology as a science. It is not a place for the fossil hunter. In other parts of the Comley area the

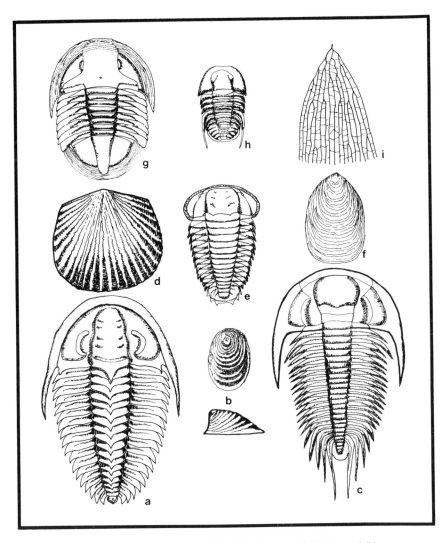

Fig. 17. Cambrian Fossils from Shropshire. Trilobites (a,c,e,g,h), Gastropod (b), Brachiopods (d,f), Graptolite (i). a. *Callavia callavei*, Lower Comley Limestones, Comley Series, x0.4; b. *Helicionella subrugosa*, Lower Comley Limestones, x1; c. *Paradoxides davidis*, St. David's Series, x0.4; d. *Orusia lenticularis*, Merioneth Series, x3; e. *Peltura scaraboides*, Merioneth Series, x3; f. *Lingulella davisi*, Tremadoc Series, x1.5; g. *Asaphellus homfrayi*, Tremadoc Series, Shineton Shales, x0.7; h. *Shumardia pusilla*, Tremadoc Series, Shineton Shales, x4; i. *Dictyonema flabelliforme*, Tremadoc Series, Shineton Shales, x0.7.

The Cambrian Period   55

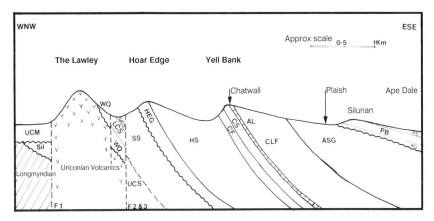

Fig. 18. Diagrammatic cross section from the Lawley to Ape Dale. Thickness of beds not to scale. UCM, Upper Coal Measures; WQ, Wrekin Quartzite; LCS, Lower Comley Sandstone; UCS, Upper Comley Sandstone; SS, Shineton Shales; HEG, Hoar Edge Grit; HS, Harnage Shales; CF, Chatwall Flags; CS, Chatwall Sandstone; AL, *alternata* Limestone; CLF, Cheney Longville Flags; ASG, Acton Scott Group; KG, Kenley Grit; PB, *Pentamerus* Beds; PS, Purple Shale; F1, F2, F3 branches of Church Stretton Fault. (After Toghill and Chell, 1984.)

overlying Upper Comley Sandstone, which rests with an unconformity on the Lower Comley Limestones and Sandstone, has yielded various Middle Cambrian fossils, particularly trilobites of the genus *Paradoxides* (Fig. 17).

Two kilometres east of the Comley area, the Geological Survey has mapped a 'complete' Cambrian sequence to the east of Hill End (Fig. 19) where the Wrekin Quartzite is well exposed, but its junction with underlying Uriconian Volcanics is not seen. The Shineton Shales are exposed but the intervening rocks are not.

One kilometre south of the Comley area is the Cwms on the south flank of Caer Caradoc. Here, Cobbold excavated sections in the Wrekin Quartzite and Comley Sandstones which showed that the former rested unconformably on reddish sandstones of Western Longmyndian age, but this exposure has always been questionable, particularly the nature of the so-called Longmyndian.

No further outcrops of Cambrian rock occur within the Church Stretton Fault system in Shropshire, but they do occur in north Herefordshire, the Malverns, the Nuneaton area, and as far south as the Bristol area. They were clearly deposited over most of the English Midlands, and in many places removed by post-Cambrian/pre-Ordovician uplift and erosion, as seen directly south of the Comley-Hill End area around Hope Bowdler, where Ordovician rocks rest with a marked unconformity on a deeply eroded Uriconian Volcanic landsurface (see later).

West of the Church Stretton Fault, the late Cambrian occurs again as a 1 km-wide band of Shineton Shales directly west of the Pontesford-Linley

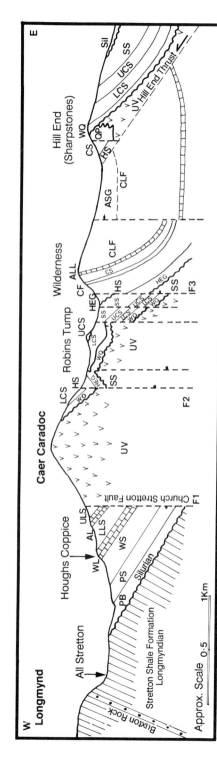

Fig. 19. Diagrammatic cross section from the Longmynd, across Caer Caradoc, to Hill End. Thickness of beds not to scale. PB, *Pentamerus* Beds; PS, Purple Shale; WS, Wenlock Shale; WL, Wenlock Limestone; LLS, Lower Ludlow Shale; AL, Aymestry Limestone; ULS, Upper Ludlow Shale; UV, Uriconian Volcanics; WQ, Wrekin Quartzite; LCS, Lower Comley Sandstone; UCS, Upper Comley Sandstone; SS, Shineton Shales; HEG, Hoar Edge Grit; HS, Harnage Shales; CF, Chatwall Flags; CS, Chatwall Sandstone; ALL, *alternata* Limestone; CLF, Cheney Longville Flags; ASG, Acton Scott Group; QP, Quartz Porphyry in UV, F1, F2 F3, branches of Church Stretton Fault. This figure clearly shows that the Church Stretton valley is not a rift valley (cf. Dineley, 1960, Fig. 5), since the main fault (F1) simply downthrows Silurian rocks to the west where they rest unconformably on Longmyndian rocks in the floor of the valley. (After Toghill and Chell, 1984.)

Fault, and all along its length (Fig. 4) from Pontesford to Linley. The shales are followed by the basal Ordovician Stiperstones Quartzite. In this western area there are no outcrops of older Cambrian rocks at all. The outcrop width of the Shineton Shales suggests a possible thickness of 1,000 m. It is difficult to imagine that the rest of the earlier Cambrian Sequence was not laid down here in west Shropshire. It was probably removed by later folding and faulting, and may be present at depth.

Chapter 5
Volcanoes, Deep and Shallow Seas — the Ordovician Period

The Ordovician period in Britain is well known for its great thicknesses of volcanic lavas and ashes, particularly in Snowdonia and the Lake District. Shropshire lay within reach of the great volcanoes of North Wales and had its own volcanic islands as well. The seas around supported a great variety of marine animals, now fossilised. The south east margin of the Iapetus Ocean was actually situated across Shropshire during the early and middle part of the Ordovician causing half the county to be land and half to be under the sea.

The history of the term Ordovician is a classic story well known to all geologists. It was first used by Charles Lapworth, first Professor of Geology at Birmingham University, in 1879, and settled (or at least started to settle) an argument which had been raging since 1840 between two of the greatest Victorian geologists, Professor Adam Sedgwick at Cambridge and Sir Roderick Impey Murchison, Director of the Geological Survey.

These two men had set out as firm friends in the 1830s to investigate the older rocks in Wales and the Welsh Borders, which no one had looked at in detail before. Sedgwick went to North Wales and Murchison to the Welsh Borders, including Shropshire. They were both true pioneers in geology and great men. Sedgwick found the going quite hard, but by the start of the 1830s he had worked out, quite correctly, the order of the rock succession in North Wales, and in 1835 named them all the Cambrian System. Murchison found the going somewhat easier and published his great work on the rocks of the Welsh Borders, *The Silurian System* in 1839, having first used the term Silurian in 1835, naming it after the ancient British tribe, the Silures, who once inhabited the area.

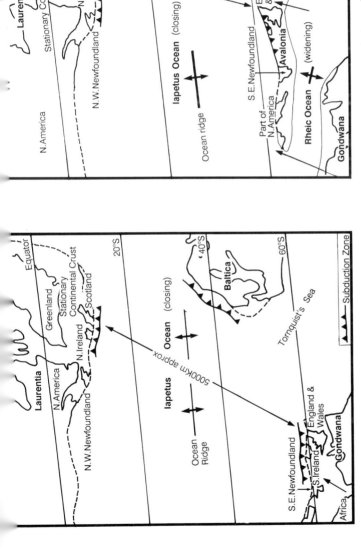

Fig. 20. Paleogeographic reconstruction of the Iapetus Ocean in the lower and upper Ordovician. During the lower Ordovician the Iapetus Ocean starts to close as Gondwana, together with southern Britain and south east Newfoundland, starts to move north. Subduction zones form on either side of the ocean as shown. At this stage Laurentia is stationary, and Baltica moving north very slowly. During the middle and upper Ordovician the Rheic Ocean starts to form, causing what is now called Avalonia (southern Britain and south east Newfoundland) to break away from Gondwana and move north towards Baltica, causing Tornquist's Sea to close by the end of the Ordovician. The collision of Avalonia with Baltica in the late Ordovician (Ashgill) causes the major Shelveian (Taconian) orogeny over southern Britain. This one mass (Avalonia and Baltica) continues to move north during the Silurian, with the Rheic Ocean opening to the south, but does not finally collide with Laurentia (including Scotland) and close the Iapetus Ocean until the end of the Silurian, with the main orogenic episode of southern Britain occurring in the mid-Devonian as the Acadian (Late Caledonian movements). During all this time Laurentia is stationary over the equator and only the Avalonian microcontinent and Baltica move north. See Fig. 33 for the situation in the early Silurian. (After Cocks and Fortey, 1982; Anderton, et al., 1982; Cocks and McKerrow, 1986).

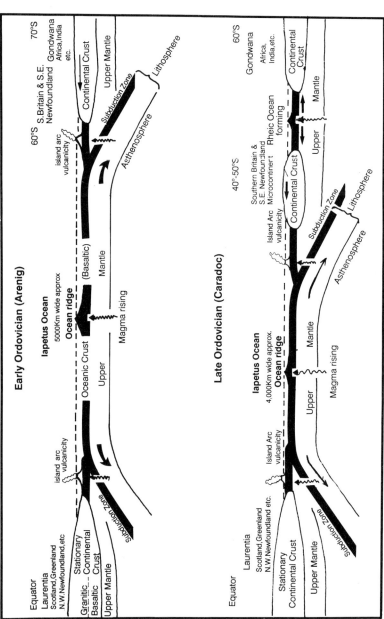

Fig. 21. Cross section of the Ordovician closing of the Iapetus Ocean. Gondwanaland, including England and Wales, and south east Newfoundland (Fig. 20) started to move north in the lower Ordovician (as did Baltica) towards Laurentia (including Scotland) which remained stationary near to the equator. Subduction zones formed on both sides of the ocean in the early Ordovician (possibly during the late Cambrian on the Laurentian margin) and were well established by the middle Ordovician. Both these subduction zones led to widespread island arc vulcanicity. During the mid-Ordovician the southern Britain-south east Newfoundland microcontinent broke away from Gondwana as the Rheic Ocean formed in between. This microcontinent moved quickly north, with both subduction zones still operating. During the Ordovician, Gondwana was

When the fossils of the two great systems were compared it was found that a great deal of Murchison's Lower Silurian covered the same time interval at Sedgwick's Upper Cambrian — the rocks were different but the fossils were the same. Murchison, politically the stronger of the two, claimed most of Sedgwick's Cambrian as Silurian. Sedgwick, and particularly his supporters, claimed the rocks he had described should be Cambrian, and that Murchison should give way. Charles Lapworth, in looking at the rocks of Wales, the Welsh Borders and South Scotland, suggested that in fact three great faunal (fossil) divisions were present, the true criteria for erecting geological periods and systems. The lowest, he said, should be the Cambrian System, the upper the Silurian System, and the middle, which covered the overlapping controversial sequence of rocks and fossils, should be named the Ordovician System, after the ancient British tribe, the Ordovices, who inhabited parts of North Wales. By this time (1879), both Sedgwick and Murchison were dead but the argument lingered on until the term Ordovician appeared on Geological Survey maps in the early 1900s. A permanent effect of the controversy still noticeable today is that the Geological Survey, when labelling their maps in the last century, used the letter 'a' for Cambrian, 'b' for Silurian, 'c' for Devonian, etc., and this is still done today. When the Ordovician System was accepted, of course there was no appropriate letter between 'a' and 'b' to use, and so even today 'b' covers both Ordovician and Silurian rocks on all Geological Survey maps.

During the Ordovician period 490-435 million years ago, the Iapetus Ocean continued to close over the British area with subduction zones on both sides (Figs. 20 and 21). These led to deep-seated igneous, volcanic and earthquake activity on both margins, and in southern Britain this was continuous throughout the lower and middle part of the Ordovician, dying away in the later part. The volcanic activity was often of the island arc type. Major earth movements (an orogeny) giving rise to fold mountains and intense deformation (folding and faulting) of strata, as well as the formation of metamorphic rocks and emplacement of deep-seated igneous rocks (plutons), commenced in Scotland in Ordovician times as an early stage of the Caledonian orogeny, which reached its climax over Scotland in the late Silurian. However, the climax in southern Britain occurred during the Devonian, but an important period of earth movements, often referred to as the Taconian, affected Shropshire in the late Ordovician.

The term Taconian comes from North America and is used to cover late Ordovician events there, and may not be an appropriate term to use in southern Britain. Important folding and faulting occurred in southern Britain, associated with igneous activity, caused by the collision of southern Britain and south east Newfoundland (Avalonia) with Baltica during the Ashgill (late Ordovician), as Tornquist's Sea (Fig. 20) closed. A better name is needed for these most important earth movements, and I suggested earlier that the term 'Shelveian' orogeny would be appropriate, since many of the structures, such as the Shelve anticline occur in the Shelve area. The name Caledonian should perhaps be restricted only to the orogeny which affected the Laurentian margin, e.g. Scotland, which culminated in the late Silurian. In southern Britain, including Shropshire, the next major orogeny is of mid-Devonian age, caused when Avalonia and Baltica collided with Laurentia, and here the term Acadian is perhaps a better term than Caledonian, and is already used in North America for early to mid-Devonian events. Caledonian suggests late Silurian earth movements, but in Shropshire there is no break in the stratigraphic sequence from the base of the Silurian (post Shelveian orogeny) until the base of the Upper Old Red Sandstone (Upper Devonian). Sedimentation was continuous from the early Llandovery right through the Silurian into the Devonian (Old Red Sandstone) to the top of the Lower Old Red Sandstone. Only a change from marine to non-marine rocks in the late Silurian marks the time of the Caledonian orogeny further north. The absence of the Middle Old Red Sandstone in Shropshire indicates the time of the Acadian orogeny, as dated by the intrusion of the Shap granite in the Lake District, which is lower Devonian, dated at 393 million years old.

Invertebrate marine life was evolving rapidly in the Ordovician, which is well known for its beautiful fossils, but there was still no life on land, nor had any vertebrates in the form of early fish appeared in the seas, although there are obscure records from North America.

The Ordovician System is divided into five smaller divisions termed Series. These are, in descending order:

Ashgill Series
Caradoc Series
Llandeilo Series
Llanvirn Series
Arenig Series

It should be said here that most countries include the Tremadoc Series of the British late Cambrian as the basal series of the Ordovician, and British geologists, in particular those of the British Geological Survey, are alone in considering the Tremadoc as latest Cambrian.

Each of these time divisions is named after an area where the rocks of that time were first studied, and are best exposed. This area is what geologists called a type area. The Caradoc Series is named after the area east of Caer Caradoc where the rocks of that age are well exposed and were first studied in detail by Murchison, who actually included it in his lower Silurian. The other names are places in Wales, and one in the Lake District. It must be said that type areas quite often turn out nowadays to be *not* necessarily the best place to study a particular age of rocks. Although the type area was by definition the first to be studied, it may be that a better area is found some years later exposing rocks of the same age. Thus the Arenig area in Wales has now been found to have a number of gaps in its 'type' sequence and another area in South Wales has now been chosen as the type area. Another difficulty comes with variations in rock types. The Caradoc area of Shropshire contains a highly fossiliferous shallow water shelf facies, and is difficult to compare with areas of deeper water sedimentation in Wales with a fauna of graptolites, or graptolitic facies. The word facies is another one of these terms which all geologists understand but is difficult to explain in a few words. Facies is the sum total of features such as sedimentary rock type, mineral content, fossil content, which characterise a sediment as having been deposited in a given environment. Thus reef facies indicates limestone in a reef environment. Graptolitic facies suggests deep water sediments with only graptolites as fossils. Shelf facies indicates continental shelf conditions. The term is also used for certain metamorphic conditions of temperature and pressure.

The Ordovician rocks of Shropshire

The Ordovician period is one of the most interesting in Shropshire because of the variety of rock types, and because the rock sequence is very different east and west of the Pontesford-Linley Fault, which during the lower and middle part of the period marked the actual shoreline of the south east margin of the Iapetus Ocean in Shropshire and further south west (Figs. 22 and 23). The Ordovician rock sequence of the classic Stiperstones-Shelve area west of the Pontesbury-Linley Fault is very different from that of the type Caradoc area east of Church Stretton. The Shelve area shows a complete Ordovician sequence, often deep water and with volcanic lavas and tuffs, from the base of the Arenig Series, to the top of the Caradoc Series, whereas east of Church Stretton the sequence does not start until the Caradoc Series, no earlier Ordovician rocks being present. This area east of the Pontesford-Linley Fault was a land mass not transgressed by the sea until Caradoc times, when a shallow shelf sea (Fig. 23) transgressed from the west laying down a thin shelf shelly facies of sedimentary rocks very rich in fossils, and very different from its equivalent rock sequence in the Shelve area, which is a mainly mixed graptolitic and shelly facies.

In many books and papers, the marked facies change between the Shelve and Caradoc areas (both rock and fossil changes) is said to be either side of the Church Stretton Fault. In fact the change occurs either side of the Pontesford-Linley Fault, although the main outcrops of Caradoc sequences in the east are east of the Church Stretton Fault. The critical area for the appreciation of this change is at Pontesford itself, where a small area of

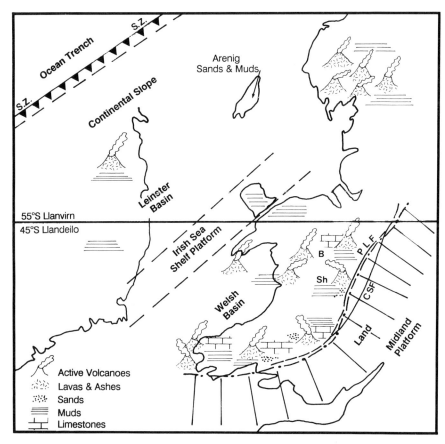

Fig. 22. Lower Ordovician (Llanvirn-Llandeilo) palaeogeography of southern Britain. Palaeolatitudes based on Cocks and Fortey (1982). PLF, Pontesford-Linley Fault; CSF, Church Stretton Fault; B, Berwyns; Sh, Shelve area; SZ, Subduction Zone. (After Toghill and Chell, 1984.)

Ordovician rocks occurs just east of, and against, the Pontesford-Linley Fault. Other important outcrops of Ordovician rocks in the county occur around the Breidden Hills (mostly in Wales) and in the far north-west of the county, west of Oswestry where Ordovician rocks, which are really part of the Welsh sequence of the Berwyn Mountains, are found.

The Stiperstones-Shelve area

This still remote area of west Shropshire (and part of Powys) is a classic area for the geologist, where the varied scenery is a reflection of the great variety of Ordovician rock types. The area also contains the remains of a

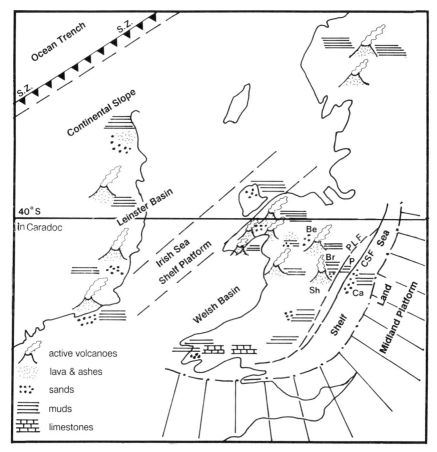

Fig. 23. Upper Ordovician (Caradoc) palaeogeography of southern Britain. PLF, Pontesford-Linley fault; CSF, Church Stretton Fault; Ca, Caradoc area; Sh, Shelve area; Br, Breiddens; Be, Berwyns; P, Pontesford; SZ, Subduction Zone. (After Toghill and Chell, 1984.)

once thriving lead and barytes mining industry based on the abundance of mineral veins which cut certain strata, but which are themselves of late Devonian age.

The area, usually known as the Shelve inlier, is directly to the west of the Pontesford-Linley Fault (Fig. 4), and the basal Ordovician beds overlie the late Cambrian (Tremadoc) Shineton Shales. Earth movements at the end of the Cambrian caused the sea to retreat to a line just west of the Pontesbury Fault so that the lower Ordovician rocks were only laid down in this area.

The Shelve area sequence (Whittard, 1979) is complete from the base of the Arenig Series to somewhere in the upper part of the Caradoc Series. The latest Ordovician Ashgill Series rocks are absent, and the basal

66   *Geology in Shropshire*

Silurian rocks rest with a marked angular unconformity on the Ordovician rocks of the area all around the north, south and west sides. The types of sedimentary rocks found suggest alterations of shallow and deeper water conditions and the fossils (Fig. 24) reflect this in that certain beds are full of planktonic graptolites, whereas others are full of shallow water brachiopods, and trilobites are common in many of the rock formations. The fauna is often referred to as a mixed shelly and graptolitic facies.

The presence of volcanic lavas and tuffs, particularly those of Llanvirn and Caradoc age associated with marine sediments, suggests that volcanic islands were present as part of an island arc system associated with a major

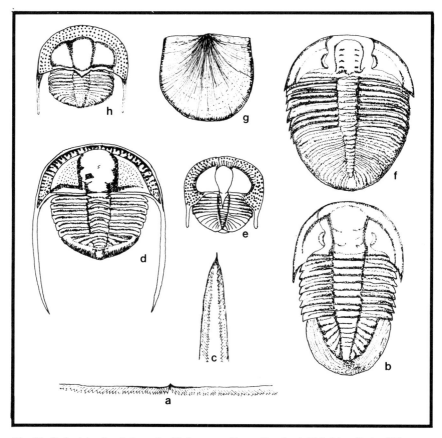

Fig. 24. Ordovician fossils from the Shelve area. Graptolites (a,c), Trilobites (b,d,e,f,h), Brachiopod (g). a. *Didymograptus hirundo*, Mytton Flags, Arenig Series, x0.7; b. *Ogygiacaris selwyni*, Mytton Flags, Arenig Series, x0.7; c. *Didymograptus murchisoni*, Betton Beds, Llanvirn Series, x0.7; d. *Stapeleyella inconstans*, Stapeley Volcanics, Llanvirn Series, x2; e. *Lloydolithus lloydiii*, Meadowtown Beds, Llandeilo Series, x1; f. *Ogygiacarella debuchi*, Meadowtown Beds, Llandeilo Series, x0.5; g. *Macrocoelia expansa*, Aldress Shales, Caradoc Series, x1; h. *Salterolithus caractaci*, Whittery Shales, Caradoc Series, x1.

subduction zone north west of Wales. Some of the volcanic ash could have travelled from Wales, but most was probably local. The volcanic island arc can be compared with those today forming the Aleutian Islands and many of the East Indian archipelagos, as well as the Lesser Antilles of the Caribbean.

The complete sequence of rocks is as shown in Table 4.

The Stiperstones Quartzite is a shallow water, very pure white quartz sandstone, not a metamorphic quartzite, and thus is similar in its formation to the Wrekin Quartzite, but purer in quartz. It contains conglomerates and thin shales, and rests on the Tremadoc Shineton Shales. The actual junction is quite sharp and as there is no discordance in dip, the junction is referred to as a disconformity, since part of the Shineton Shales faunal sequence is missing. This situation, when more local and on a smaller scale,

Table 4. The Ordovician sequence in the Shelve area.

Caradoc Series	Chirbury Formation	Whittery Shales	305 m+
		Whittery Volcanics	90 m
		Hagley Shales	305 m
		Hagley Volcanics	107 m
		Aldress Shales	305 m
		Spy Wood Grit	90 m
Llandeilo Series	Middleton Formation	Rorrington Shales	305 m
		Meadowtown Beds	400 m
		Betton Beds	180 m
		Weston Beds	?430 m
Llanvirn Series	Shelve Formation	Stapeley Shales	?240 m
		Stapeley Volcanics	?430 m
		Hope Shales	240 m
Arenig Series		Mytton Flags	910 m
		Stiperstones Quartzite	120 m
Tremadoc Series (Cambrian)	disconformity	Shineton Shales	1,000 m?

would be called a non-sequence, as in the Cambrian around Comley. The main mass of quartzite forms the spectacular ridge of the Stiperstones (Plate 15) with its well known, frost-shattered tors formed in a periglacial climate about 20,000 years ago. Fossils are virtually absent from the Stiperstones Quartzite accept for worm borings, which are quite commonly found as narrow tubes (now filled in) of up to 5 mm in diameter at right angles to the bedding. The openings to these can often be seen on the tops of bedding planes. Rare trilobite remains suggest faunal similarities with rocks now found in Brittany and other areas which were on the south side of the Iapetus Ocean. The succeeding Mytton Flags are the thickest unit of the Shelve succession, 900 m of blue-grey, flaggy greywacke siltstones, and contain both trilobites and graptolites, as well as the earliest gastropods.

Plate 15. The frost shattered tors of the Stiperstones. The Ordovician Stiperstones Quartzite dips steeply to the west (left) at about seventy degrees.

Both the Stiperstones Quartzite (Plate 16) and the Mytton Flags are sedimentary rocks showing conspicuous joints, a feature of hard sedimentary rocks and igneous rocks. When sediment is deposited in water it is wet, and in drying out under compaction, cracks develop just as they do in drying mud in a field. In sandstones and limestones the major joints are very regular — two sets of plane surfaces at right angles to the bedding and each other (Fig. 25). Major joints are usually 0.5 to 3 m apart and help form a rock which is a good building stone, in that large uniform blocks can be extracted, in which case it is called a freestone. In limestones the joints are often enlarged by solution to form fissures known as grykes. Major joints in all sedimentary and igneous rocks can be useful lines of weakness for mineral veins to follow, and in the Mytton Flags this is a very important feature as will be explained below.

The Mytton Flags pass up into finer grained black Hope Shales, suggesting deep water, with a mixed graptolitic and shelly fauna. Here the first volcanic ashes appear, termed chinastones around Hope, and these tuffs become the dominant lithology in the succeeding Stapeley Volcanics, comprising massive water-lain andestic tuffs, with some lavas, and interbedded shales, possibly erupted from small volcanic islands as the Iapetus Ocean closed, and well exposed on Stapeley Hill. The overlying Stapeley Shales pass up into the coarser Weston Beds which contain sandstones, shales and tuffs. The first abundant bivalve molluscs (ancestors

Two sets of major joints at right angles to each other, and always at right angles to the bedding. Large plane surfaces can be joint faces or bedding planes. Bedding plane traces (BPT) will appear on joint surfaces, and major joints traces (MJ) will appear at bedding planes.

a) In a steeply dipping sequence such as the Stiperstones Quartzite.

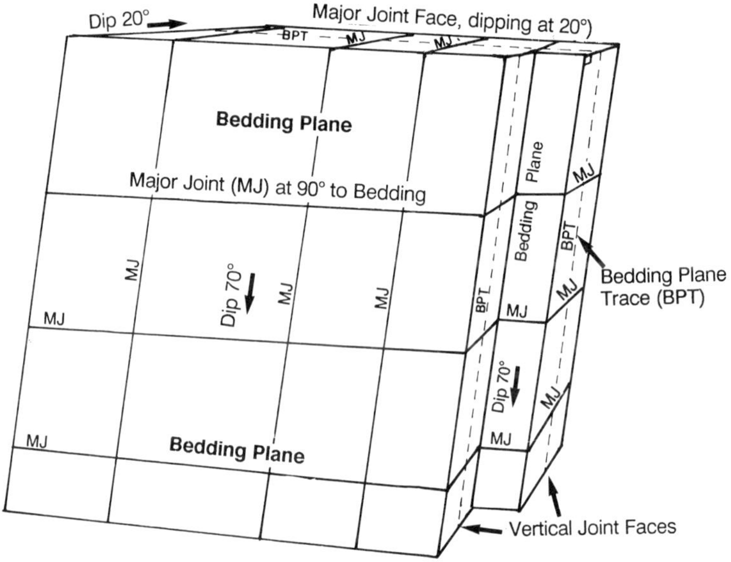

b) In a gently dipping sequence such as the Grinshill Sandstone

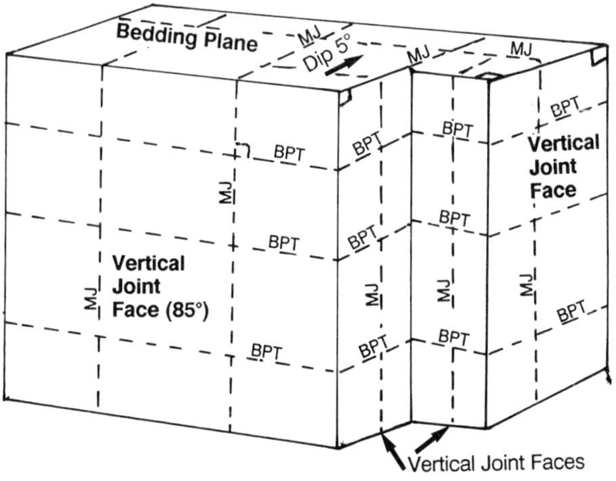

Fig. 25. Relationships between joints and bedding planes.

Plate 16. A close-up of steeply dipping Stiperstones Quartzite. Note two sets of major joints at right angles to each other and to the bedding. The dip is seventy degrees to the west (left). This relationship is as explained in text Fig. 25.

of mussels) appear in the Weston Beds, probably indicating shallow water. The Betton Beds are blue-grey shales and flags, and grade up into the Meadowtown Beds of Llandeilo age which comprise limestones, flags and tuffs, with a rich shelly fauna of brachiopods, and abundant trilobites. The overlying Rorrington Shales show a return to more graptolitic black shales, whereas the succeeding Spy Wood Grit (sandstone and shale) contains a shelly fauna which allows correlation with the Hoar Edge Grit of the Caradoc area. The remainder of the succession (listed above) comprises alternations of shales and tuffs. The Whittery Volcanics contain some very coarse agglomerates that could represent volcanic vents.

Folding and Faulting in the Shelve area
Later Ordovician rocks of Ashgill Series age are absent from Shropshire (except in the far north-west in the Berwyn Dome area), and this was a time of uplift, folding and erosion known as the Shelveian (Taconian) earth movements. These folded the Shelve area rocks into two major folds trending north east-south west — a downfold or syncline, the Ritton Castle syncline, and a corresponding upfold anticline, the Shelve anticline (Figs. 5 and 26). As well as these major folds, the whole area is affected by considerable faulting, particularly horizontal tear faults which offset

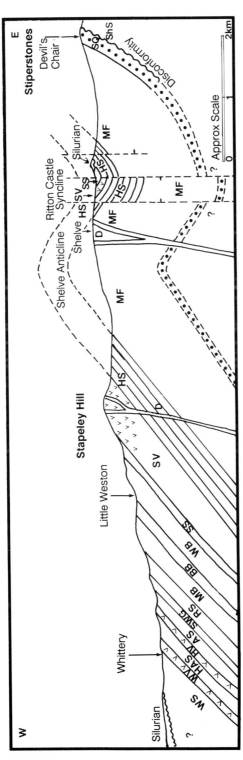

Fig. 26. Cross section through the Ordovician rocks of the Shelve Area. Thickness of beds not to scale. West of Stapeley Hill the section is diagrammatic. Information from map in Whittard (compiled by Dean) (1979). WS, Whittery Shales; WV, Whittery Volcanics; HAS, Hagley Shales; HV, Hagley Volcanics; AS, Aldress Shales; SWG, Spy Wood Grit; RS, Rorrington Shales; MB, Meadowtown Beds; BB, Betton Beds; WB, Weston Beds; SS, Stapeley Shales; SV, Stapeley Volcanics; HS, Hope Shales; MF, Mytton Flags; SQ, Stiperstones Quartzite; ShS, Shineton Shales; D. Dolerite. (After Toghill and Chell, 1984.)

outcrops for considerable distances. None of these faults cut the Pontesford-Linley Fault, which is clearly a different type and separated the Shelve area from the small Ordovician outcrop at Pontesford which has close affinities with the Ordovician sequence of the type Caradoc area east of Church Stretton. This major fault had vertical movement along it during the Ordovician to keep the eastern areas as a land mass in the early and middle part of the Ordovician, but it also probably had a good deal of lateral horizontal movement along it, and could have been a major tear fault of Great Glen Fault proportions. The Shelve Ordovician rocks of Caradoc age are very different from those of the type Caradoc area and the Pontesford area, and the early Ordovician of Shelve contains some quite deep water rock types, such as the Hope Shales. This suggests that the whole Shelve sequence (Woodcock and Gibbons, 1988) may have

Plate 18. Barytes vein on Cothercott Hill, northern Longmynd. Thickness 0.3 m.

accumulated a long way from where it is now situated, and was brought into its present position by tear faulting along the Pontesford-Linley Fault Woodcock (1984, 1988) has called this the Pontesford Lineament. The concept of large areas, nowadays called terranes, being moved large distances along major tear faults during the plate tectonics process, is being used increasingly today to explain the position of certain exotic rock sequences which seem out of place compared with surrounding rock

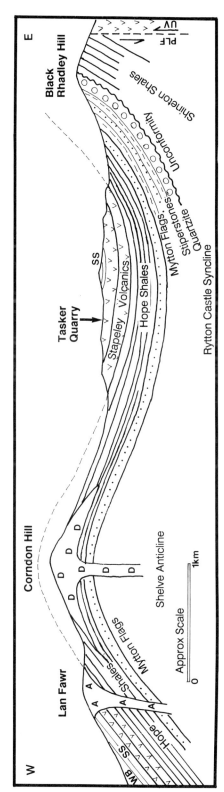

Fig. 27. Diagrammatic cross section from Corndon Hill to Black Rhadley Hill showing the Corndon dolerite intruded as a phacolith into the pre-existing fold of the Shelve anticline. WB, Weston Beds; SS, Stapeley Shales; A, Andesite; D, Dolerite; PLF, Pontesford-Linley Fault; UV, Uriconian Volcanics.

sequences. The terrane concept is being applied to many parts of western North America, where parts of California may have originated in the Gulf of Mexico. We have earlier suggested that the Longmyndian may be an exotic terrane, and the geology of many other deformed parts of Britain, particularly in Scotland, could be explained using this concept.

Intrusive Igneous Rocks
A number of large and small intrusive masses of andesite, dolerite and coarser gabbro occur. Many of the larger masses are steep-sided intrusions which could be called bosses, such as the gabbro of Squilver (Disgwylfa) Hill, and a number of dolerite dykes also occur throughout the area. Perhaps the most well known intrusion is the dolerite of Corndon Hill (Plate 17) (Fig. 4). This is a structure called a phacolith, an intrusion into an already formed anticline, in this case the Shelve anticline in the Hope Shales (Fig. 27). The hill has a lovely conical shape when viewed from the north-east. Associated with the Corndon dolerite is a rock type forming a small outcrop on the south side of the hill at Cwm Mawr near Hyssington. This is called a picrite, and the chemistry is described as ultrabasic, low in silica. It is a medium grained, blue-grey rock with breaks with sharp edges, and the locality was a stone axe factory in prehistoric times, producing axes with a unique rock type which spread far afield. South west of Corndon

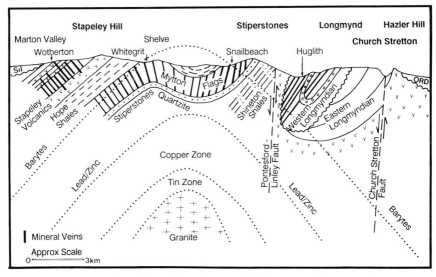

Fig. 28. Mineral zones in the Shelve-West Longmynd mining areas. Diagrammatic cross section showing zonation of minerals by distance from a buried granite source. Quartz and Calcite are the main gangue (non-economic) minerals. No Fluospar occurs, in marked contrast to the Pennine lead ore fields. Note how Snailbeach lies on the barytes/lead-zinc zone boundary and thus produced both mineral assemblages.

Hill, the hills of Roundtain, Todleth and Lan Fawr are large masses of andesite mapped as intrusions but which contain coarse tuffs and lavas erupted from volcanic islands in the area. All the intrusions in the Shelve area are latest Ordovician in age, after the Shelveian folding.

Mineralisation of the Shelve Area

Lead-zinc and barytes mineralisation is a well-known feature of the Shelve area, which gave rise to a thriving lead and barytes mining industry in the nineteenth and early twentieth centuries. It is remarkable that such a small ore field could produce ten per cent of the nation's lead ore in 1872, producing 5,000 tons of refined lead in that year, only exceeded by the larger areas of Northumberland, Derbyshire and Montgomeryshire.

The minerals occur as veins (Plate 18), thin vertical sheets of igneous material injected into the surrounding rocks at a late stage of igneous activity. As a good deal of water is involved in the process, the veins are called hydrothermal. Most were deposited from vapours and solutions rather than from pure melts. The veins can be anything from 1 cm to 3-5 m across, and represent the latest stage in cooling of an igneous body at depth, in which the concentration of rare minerals with low melting points is finally injected into the surrounding (country) rocks. This suggests that under the Shelve area at depth is a large igneous body, probably a granite, which was the source for the minerals. This has not been discovered, but is certainly there, and a deep borehole might find it. In Cornwall, the veins of tin are associated with large granite intrusions, but here erosion has actually worn down the rocks to such a depth that the parent granite is exposed. In Shropshire it may be 5-10 km under the surface, since we know that veins are roughly zoned chemically away from a parent granite as the temperature decreases. Tin veins are nearest, then copper, then zinc, lead, and finally barytes (Fig. 28).

The main minerals in the area are:

Galena — Lead sulphide, PbS. A silvery-blue, heavy, metallic-looking mineral, with a very obvious cubic cleavage, that is, when struck, the mineral breaks up into smaller and smaller cubes. Some of the galena contains small amounts of silver and is called argentiferous galena. Although there is no easy way of telling whether silver is present, the Romans certainly obtained it from here, and in 1872 Shropshire produced 3,000 oz of silver.

Sphalerite — Zinc sulphide, ZnS, sometimes called blende or zinc blende. A duller, reddish-brown to black mineral usually occurring as non-crystallised masses, but occasionally as single, many-faced crystals (tetrahedra and rhomdodecahedra) with slightly curved faces. There was never a great trade in zinc from the area. In the late 1980s, a search was carried out for cadmium which can make up to five per cent of the content of blende, but the results did not warrant mining applications.

Barytes — Barium sulphate, $BaSO_4$, sometimes called heavy spar. Very heavy, dense mineral, which reflects X-rays and is the basis of barium meals. White or pinkish white, occasionally colourless, usually occurring as large masses without crystal shape, but with two good cleavages at right angles. Nowadays it is used extensively as the basis of a dense drilling mud, and has also been used as a pigment, and to add weight to paper.

These three minerals are found in association with calcite (calcium carbonate, $Ca\ CO_3$) and quartz (silicon dioxide, silica, SiO_2). These two minerals were referred to by miners as gangue minerals, in that they were of no economic use, and were left as spoil heaps. In 1989 concern was expressed by Shropshire County Council about possible dangers from lead pollution around the large white tip at Snailbeach. However, research so far would suggest there is minimal risk to the local population. Nowadays, the calcite spoil heaps are exploited for local pebble dash, but it is a rather sharp mineral, and easily breaks. Calcite is white and can be confused with barytes in the same vein, but is much lighter in weight (less dense) and also has three perfect cleavages at 120° giving rise to perfectly formed rhombohedra when broken into cleavage fragments. Perfectly clear calcite cleavage fragments known as Iceland Spar show the strong double refraction of light, and were used in early petrological microscopes as the so-called Nicol prisms to produce polarised light.

A number of other rarer minerals occur in the veins, including iron and copper pyrite (fools gold), witherite (barium carbonate), calamine (white-green zinc carbonate), cerussite (white lead carbonate) and green pyromorphite (chlorophosphate of lead). Clots of still malleable natural pitch occur quite commonly within calcite crystals and also coating other minerals. This black, naturally occurring, solid hydrocarbon has probably been produced by downward passage of hydrocarbons from once overlying Coal Measures, now eroded away, but still found nearby at Pontesford. Or it could be a magmatic product from depth. In complete contrast to the Pennine lead ore fields no fluospar occurs in Shropshire.

Radiometric dates for the veins suggest they are late Devonian or early Carboniferous, but some writers have suggested they are late Ordovician. The veins certainly do not cut the overlying Silurian rocks at all, which might suggest they are pre-Silurian. On the other hand, even within the Ordovician rocks the veins are only found in the Mytton Flags and the harder Caradoc Series sandstones and volcanics in the west of the area. The veins follow major joints and faults in the Mytton Flags, and softer beds such as the Hope Shales, which are without major joints are devoid of veins, as are other rock types. The chemistry of the rock types through which the hot material passed was obviously very important in determining whether minerals precipitated out or not. No veins occur in the Stiperstones Quartzite and yet this is a massive sandstone with beautiful regular joint patterns (Fig. 25). Any mine following a vein in the Mytton Flags simply found that it stopped when it reached the Quartzite.

The zonation of minerals referred to above stated that the barytes zone is further away from the parent granite than the lead/zinc zone. This is true in this area and Fig. 28 shows how the curved zone boundary passes through the area twice, giving lead/zinc in the central area, and barytes in the far west, and also in the east where the northern outcrop of the Longmyndian has extensive mineralization. The Snailbeach mine appears to be actually on the zone boundary, mining barytes at shallow depths and lead deeper down. It produced 132,000 tons of lead ore between 1845 and 1913. Wotherton mine in the far west of the area, in the barytes zone, produced 125,000 tons of barytes between 1865 and 1911.

The barytes mines of the northern Longmyndian, which contain little or no galena, do contain small amounts of copper (chalcocite, copper pyrite, bornite, and secondary malachite and azurite). Although a number of mines are referred to locally as copper mines, such as at Westcott, there are no records of the tonnage produced and most of these were, like Huglith mine, large producers of barytes right up until the Second World War. The latter produced 20,000 tons of barytes per year in the mid 1930s.

The Caradoc area, Ordovician rocks east of Church Stretton

A marked change in the Ordovician of Shropshire occurs east of the Pontesford-Linley Fault, where we find only rocks of Caradoc Series age. The older Ordovician rocks were not laid down in this area, which was a land mass until the sea transgressed from the west in Caradoc times (Fig. 23). The main outcrops are within, and east of, the Church Stretton Fault System and this is the type area for the study of the Caradoc Series in the world. The large outcrop extends southwards from Harnage in the north, passing east of Caer Caradoc and the Stretton Hills and across the Onny Valley. The outcrop is split in two by Precambrian area around Cardington Hill. In this large area the generalised sequence of rocks is as follows:

	Onny Shales	0-120 m
Caradoc Series	Acton Scott Group	60-180 m
	Cheney Longville Flags	80-240 m
	alternata Limestone	0- 20 m
	Chatwall Sandstone	40-160 m
	Chatwall Flags	30-100 m

78 *Geology in Shropshire*

          Harnage Shales   100-300 m
          Hoar Edge Grit   0-120 m

There are marked variations in thicknesses along the outcrop, and in particular the Onny Shales are only found south of Church Stretton. The landscape over which the shallow sea spread from the west was deeply eroded and irregular, and made up of Cambrian and Precambrian rocks, so that the basal Hoar Edge Grit rests with a marked unconformity on rocks ranging in age from Longmyndian in the Onny Valley (Fig. 29) to late Cambrian Shineton Shales at Hoar Edge itself (Fig. 18). Around Hope Bowdler (Fig. 30) the Hoar Edge Grit is missing, having been overstepped by the Harnage Shales which rest here with a marked unconformity on an irregular eroded surface of Uriconian Volcanics. The classic roadside exposure at Hope Bowdler (Fig. 30) shows how irregular this surface was, with Harnage Shales covering an obvious old rocky shoreline. People who visit this site should simply look at it and not hammer it, since it is impossible to collect a sample of an unconformity, which is a structural relationship. One kilometre to the west on Hazler Hill is a small quarry usually referred to as Hazler Quarry which exposes a steeply dipping sequence of Uriconian lavas and tuffs 15 m thick. In this quarry (Fig. 30) is a narrow fissure cut down about 3 m into the volcanics which contains Ordovician fossiliferous sandstone. This is called a neptunean dyke and, clearly, the Precambrian volcanic rocks of the Ordovician shoreline area contained a number of fissures, as on any rocky shoreline today. As the sea spread over them, the fissures filled with sand of Harnage Shales age, complete with fossils, and is still preserved today. Like the unconformity at Hope Bowdler, this site needs protection from too much hammering and collection. Neptunean dykes are rare geological features although one may well wonder why this is, as sand-filled fissures are so common on modern rocky shorelines.

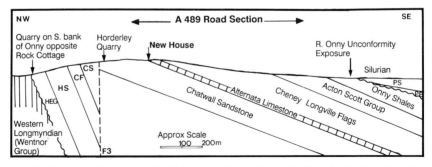

Fig. 29. Diagrammatic cross section through the Ordovician rocks (Caradoc Series) along the Onny Valley. Thickness of beds not to scale. HEG, Hoar Edge Grit; HS, Harnage Shales; CF, Chatwall Flags; CS, Chatwall Sandstone; PB, *Pentamerus* Beds; PS, Purple Shale; F3, branch of Church Stretton Fault. (After Toghill and Chell, 1984.)

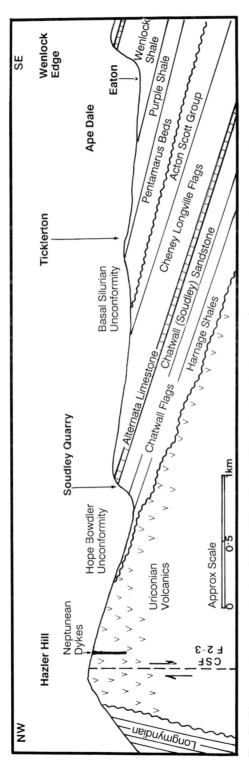

Fig. 30. Diagrammatic cross section from Hazler Hill to Wenlock Edge showing: the Neptunean Dykes on Hazler Hill (of late Ordovician, Caradoc Series, age); the famous Ordovician/Precambrian unconformity at Hope Bowdler; and the Ordovician sequence around Soudley of Caradoc Series age.

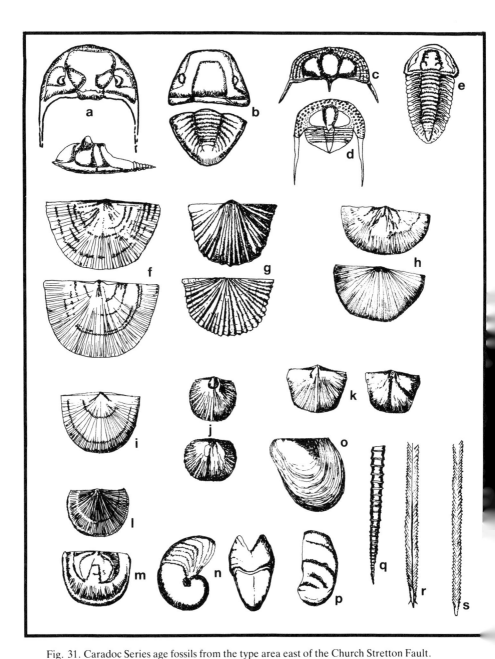

Fig. 31. Caradoc Series age fossils from the type area east of the Church Stretton Fault. Trilobites (a-e), Brachiopods (f-m), Gastropod (n), Bivalve (o), Ostracod (p), Scaphopod (q), Graptolites (r,s). a. *Chasmops extensa*, x0.4; b. *Brongniartella bisulcata*, x0.4; c. *Broeggerolithus broeggeri*, x1; d. *Onnia superba*, x0.5; e. *Flexicalymene caractaci*, x0.5; f. *Strophomena grandis*, x0.5; g. *Nicolella actoniae*, x0.75; h. *Sowerbyella sericea*, x1; i. *Kjaerina bipartita*, x0.5; j. *Dalmanella horderleyensis*, x0.75; k. *Reuschella horderleyensis*, x0.5; l,m. *Heterorthis alternata*, x0.5; n. *Sinuites*, x0.5; o. *Modiolopsis orbicularis*, x0.4; p. *Tallinnella scripta*, x5; q. *Tentaculites*, x1.25; r. *Diplograptus multidens*, x0.5; s. *Orthographtus truncatus*, x0.5.

The Caradoc area sequence is one of shallow water sedimentary rocks, sandstones, shales, flags and thin limestones, all with a very rich fauna of shallow water (benthonic) fossils, brachiopods and trilobites in particular (Fig. 31). Some graptolites occur and beds such as the *alternata* Limestone are literally fossilised shell banks crowded with one particular brachiopod, *Heterorthis alternata* (Fig. 31). The type area for the Caradoc Series is the famous Onny Valley, where the whole sequence is exposed. By comparing the brachiopod and graptolite faunas, it has been deduced that the Hoar Edge Grit of the Caradoc area is the same age as the Spy Wood Grit of the Shelve area, and thus the rocks of the two successions are all of comparable age. Grit is an ill-defined, but nevertheless useful, term for a coarse sandstone, such as the Millstone Grit. The Hoar Edge Grit contains at its lower levels coarse-grained sandstones, often pebbly, and containing examples of well-rounded grains and wind-faceted quartz pebbles, known as dreikanter (three edged).

These wind-blown grains and wind-worn pebbles have clearly been derived from the shoreline of an arid landscape nearby and blown into shallow water, where they formed a marine deposit with brachiopods. This tells us something of the nature of the Ordovician land surface which the late Ordovician sea transgressed over in the Caradoc area.

The Harnage Shales are still quite a shallow water deposit and contain abundant small ostracods (Fig. 31) which still live today, as different species. The coarser Chatwall Flags and Sandstone contain a very rich brachiopod and trilobite fauna, and the *alternata* Limestone, already mentioned as a fossilised shell bank, is lenticular, as you would expect. The Cheney Longville Flags contain bedding planes crowded with strange, ornamented, rod-shaped fossils, about the size of a matchstick called *Tentaculites*. These all often point in one direction and are clearly current orientated. They are related to the modern mollusc, elephant's tusk shell (Fig. 31). The term 'Flags', another ill-defined but useful term, describes well-bedded sandstones which split into slabs about 3-6 cm thick, useful for flagstones. The Acton Scott Group contains a thick, impure limestone around Acton Scott, and the Onny Shales are well known for their beautiful examples of the blind trilobite *Trinucleus* (Fig. 31).

Unlike the Shelve area, igneous rocks are unknown from the Ordovician of the Caradoc area, except for a thin basalt lava flow in the Harnage Shales at Sibdon Carwood, just west of Craven Arms. This is the only example of Ordovician vulcanicity in the area.

Ordovician Building Stones of the Caradoc area
A number of local Ordovician rocks have been used as local building stones and one in particular as a roofing tile. The Hoar Edge Grit was used extensively by the Romans at Uriconium (Wroxeter), presumably

transported from the Acton Burnell area along Watling Street, which was their main approach route from the south around Leintwardine through the Church Stretton valley. Within the Hoar Edge Grit calcareous flagstones occur in the northern area of the outcrop around Coundmoor and Harnage. These have sometimes been called the *Subquadrata* Limestone because of the abundance of the brachiopod *Orthis subquadrata*. They split into thin slabs suitable for roof tiles and many of the older houses in the area were roofed in this way, although few remain today. Locally the stone is called the Harnage Slate, but this is a complete misnomer. The hand-cut tiles are very heavy and the roof timbers needed to support them had to be very substantial. The calcareous flagstones do not occur south of Lodge Hill, and thus are absent on Hoar Edge itself where massive yellow white sandstones predominate.

The Chatwall Sandstone is the best known local building stone, beautifully banded purple, brown and green. It is variously called the Soudley or Horderley Sandstone in the area south of Chatwall itself. Almost all the older, small houses, and many of the larger ones, and churches, in the area around Church Stretton and to the east are built of this stone, from Chatwall in the north as far south as Cheney Longville. Around Acton Scott a thick development of impure Limestone, the Acton Scott Limestone, is used as very local building stone and is a pleasant creamy brown-yellow colour, often full of fossils.

Folding and Faulting (Tectonics) in the Caradoc area
Latest Ordovician Ashgill Series rocks are absent in this part of Shropshire, as at Shelve, and the period closed with the Shelveian (Taconian) earth movements. All the Caradoc area rocks were tilted to give them a gentle dip to the south-east (Fig. 30), but this increases near to the Church Stretton Fault which was active at the time. Here the dips are sometimes vertical, as in the Onny Valley (Fig. 29). The Ordovician strata are composed of alternations of hard and soft rocks and where the dips are gentle this has resulted in the formation of conspicious scarp and vale scenery, the harder rocks forming the escarpments such as Hoar Edge and Yell Bank, with the soft shales forming the vales in between (Fig. 18). Because the rocks dip south-east, the escarpments face north-west. Chapter 6 gives a fuller explanation of the formation of escarpments as shown around Wenlock Edge.

Folding and complex faulting of late Ordovician age is present within the Church Stretton Fault system, and vertical movements along it and the Pontesford-Linley Fault, helped to keep this eastern area above sea level in the early and middle Ordovician. As with the Pontesford-Linley Fault, there was probably a good deal of horizontal tear faulting as well, and a number of low angle faults called thrusts occur in the Hill End area

(Fig.19). All these movements occurred before the beginning of the Silurian period, and they caused the sea to be pushed back westwards to a north-south shoreline running from the Oswestry area to Welshpool at the end of the Caradoc. The Silurian rocks are not affected by this period of late Ordovician folding and uplift, and erosion has resulted in the Silurian rocks resting with a marked unconformity on various Ordovician rocks in the Caradoc and Shelve areas.

The Pontesford area

This critical locality close to Pontesbury (Fig. 4) lies 1.2 km east of the northern end of the Shelve area and contains a small isolated outcrop of Ordovician rocks, the Pontesford Shales, on the north east side of Earl's and Pontesford Hills. Here, just to the east of the Pontesford-Linley Fault, Dean and Dineley (1961) described a basal conglomerate full of Uriconian Volcanics resting on western Longmyndian rocks. This is followed by 300 m of shales and mudstones with a graptolite and shelly fauna which compares directly with the Harnage Shales of the Caradoc area, as does the lithology. Because of this similarity between the rocks and fauna, the area is quite clearly part of the Caradoc area sequence and cannot be compared with the much closer Shelve sequence. This locality proves that the Pontesford-Linley Fault marks the change between the two rock and faunal facies of the Shelve and Caradoc areas, and the change takes place over a remarkably small distance, with the Pontesford-Linley Fault marking the shoreline in the early and middle (pre-Caradoc Series) Ordovician (Fig. 22).

The area west of Oswestry

A small part of north west Shropshire (Figs. 4 and 5) includes the eastern fringe of the Berwyn Dome, where a thick sequence of Ordovician sedimentary rocks and tuffs passes under the Carboniferous Limestone. The main fold structure is an elongated anticline that dips away in all directions, like an elongated upturned saucer, hence the term 'dome'. The eastern part of this dome is covered by the Carboniferous rocks of the Oswestry area — the southern end of the North Wales Coalfield. The axis of the main fold strikes east under the Carboniferous, and the Ordovician rock sequence is repeated on the northern and southern limbs of the fold. The succession worked out by the Geological Survey, and not looked at in detail for many years is as follows:

84 Geology in Shropshire

Ashgill Series	Shales
Caradoc Series	Bryn Beds (Shales) Teirw Beds (Slates and sandstones) Cwm Clwyd Ash
Llandeilo Series	Mynydd Tarw Group (Shales and mudstones, ashes at base not seen in Shropshire)

Thicknesses are difficult to estimate, as the whole area in Shropshire is covered by glacial drift, but it must be at least 2,000 m. The sequence is part of the deeper water basin facies of Wales containing volcanic tuffs which can be traced westwards towards Bala, and includes the only Ashgill Series rocks found in Shropshire. Sedimentation in this area thus continued

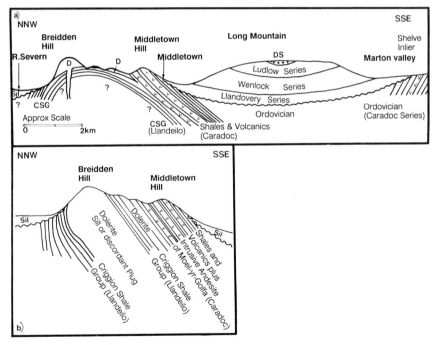

Fig. 32. Diagrammatic cross section across the Breidden Hills and Long Mountain syncline. Thickness of beds not to scale. Sil, Silurian; DS, 'Downton' Series; D. Dolerite; CSG, Criggion Shale Group. Section 'a' shows the Criggion dolerite as a laccolith intrusion (the traditional view), which caused further doming up of the already formed Breidden anticline. The domed-up beds above the dolerite have since been removed by erosion. Section 'b' shows the Criggion dolerite as a thick sill within a steeply dipping Ordovician sequence, or possibly intruded as a plug. Neither possible fold structure or type of intrusion has been proved with any certainty.

right through to the end of the Ordovician, and into the Silurian, with little break.

The sequence contains a number of outcrops of fine-grained dark grey igneous rocks called keratophyres — the largest is one square kilometre — which are probably intrusions. Within these masses are found volcanic breccias (angular coarse fragmentary rocks) which probably represent the actual sites of old volcanic necks. Keratophyres are usually lavas derived from andesitic magmas (melts) and contain sodium feldspars (albite), chrolite and pyroxene, but these rocks appear to be intruded into the Ordovician sedimentary rocks, cutting across various junctions, and are thus shallow intrusions (termed hypabyssal). Lavas which did not quite reach the surface are by definition intrusions. Keratophyres are often linked to spilite lavas, which are usually submarine in origin and form pillow lavas — globular masses — when erupted into sea water. Clearly, these rocks were not erupted and may represent an old eroded vent complex. The instrusions are post-Ordovician pre-Carboniferous in age and thus could be associated with the late Silurian-early Devonian folding of the area during what is called the Caledonian orogeny, or they may represent late Ordovician igneous activity.

There is little break between the Ordovician and Silurian in this area, and the folding of the Berwyn Dome, which includes the Silurian sequence, is thus post-Silurian pre-Carboniferous. It is part of the classic Caledonian orogeny which folded up Wales, the Lake District and Scotland into a vast, high, fold mountain chain, the Caledonian Mountains, which had been formed by middle Devonian times. This area is structurally part of Wales, and the Shelveian (Taconian) earth movements mentioned in other parts of Shropshire had little effect in this area.

The Breidden Hills

The Ordovician area of the Breidden Hills (mainly in Powys) lies 14 km north west of the Shelve area and is separated from it by the long Mountain syncline occupied by Silurian rocks (Figs. 4, 5 and 32). The Ordovician rocks of the inlier are overlain unconformably by Silurian rocks on both the south east and north west sides, but the north west junction is complicated by the line of the Severn Valley Fault System which is entirely drift covered (Fig. 5). The north east margin of the inlier is faulted against Upper Coal Measures (Fig. 5).

A sequence of shales and volcanics up to 1,100 m thick intruded by andesites and dolerites, including the famous Criggion dolerite, was originally mapped by Watts in 1885, and an anticlinal structure for the area, the Breidden anticline, has long been assumed (Fig. 32). However, little modern work was done on the area until the 1980s. The sequence is fossiliferous and correlates with the upper part of the Shelve succession from the Rorrington Shales upwards, Upper Llandeilo to middle Caradoc (Soudleyan Stage) (Whittard, 1979).

Dixon (unpublished Ph.D thesis, University College of Wales, Cardiff) has recently studied the Ordovician Sequence in great detail and has very kindly, in 1986, made detailed information available to me, on which this abbreviated account is based.

Table 5. The Ordovician sequence in the Breidden Hills.

			unconformity	Silurian	
Caradoc Series	Breidden Volcanic Group		Hill Farm Formation	Upper Shale Group	
			Bulthy Formation	'Bomb' Rock	Upper Volcanic Group
			Middletown Formation	Chinastone Ash & Acid Tuffs	
			Stone House Shale Formation	Middle Shale Group	
			Newton Brook Formation	Lower Volcanic Group & Black Grit	
Llandeilo Series	Criggion Shale Group		Criggion Shale Group	Lower Shale Group	
			Dixon 1986 (unpublished)	Watts 1885, 1925	

He has worked out a new succession shown in Table 5 and correlated it with the old succession of Watts (1885, 1925).

The succession worked out by Dixon comprises andestic tuffs interbedded with dark shales and mudstones, and cut by andesitic and basaltic intrusions.

The Criggion Shale Group is black shales. The Newton Brook Formation and Stone House Formation comprise silty shales interbedded with volcanic tuffs of submarine origin. The later Middletown Formation comprises coarse water-lain volcanic breccias and finer tuffs which represent the emergence of island volcanoes, whose erupted debris covered the area and spread to the Shelve area as well. The overlying Bulthy Formation comprises water lain volcanic conglomerates, massive boulder and cobble conglomerates, the 'Bomb Rock' of earlier workers. The Hill Farm Formation marks a return to shale depositions with thin tuff horizons, and on fossil evidence correlates with the Hagley Shales of Shelve. Dixon has shown how the volcanic material was laid down in submarine fans, often as turbidites, on the flanks of the volcanoes whose peaks rose above sea level. He has also deduced that these volcanic rocks and sediments were then, while still unconsolidated and wet, intruded by the thick dome-like Moel-y-golfa andesite. Later dolerites of the Breiddens, including the Criggion dolerite, are remants of a complex of laccolithic, sills and dyke-like intrusions, of late Ordovician, pre-Silurian age. However, they could be high level intrusions of the same age as the main Breidden Volcanic Group, but being relatively dense would appear in the deepest levels of the sedimentary basin, and it is certainly true that the thickest mass, the Criggion dolerite is entirely intruded into the Criggion Shales at the base of the sequence.

The rest of this account is based on my own opinions and those of earlier workers prior to Dixon's study. The Criggion dolerite has usually in the past been regarded as a laccolith intrusion which caused further anticlinal folding of the Criggion Shales, but the anticlinal nature of the Breidden outcrop (folded at the end of the Ordovician period) is not certain, and the dolerite could be a thick sill intruded into a sequence which dips uniformly south east. The two possible structures are shown in Fig. 32. During quarry investigations in 1980, 274 m of dolerite were penetrated under Rodney's Pillar on Breidden Hill, and the base not reached. The intrusion could be a discordant plug as the shales on the western contact do dip steeply west in places.

The Criggion dolerite is a well known local road stone with a mottled green-blue appearance. It is a medium- to coarse-grained dolerite with plagioclase feldspar making up thirty-five to sixty per cent of the rock and green pyroxene, epidote and small amounts of olivine (up to ten per cent). The quarries on the north west side of Breidden Hill are very large and proceed in a series of steps to within a few metres of Rodney's Pillar on the very top of the hill. Because of the steepness of the hill, and the difficulties in maintaining haul roads to the top of the hill, the company working the quarry, ARC, created a novel scheme in 1979 whereby a near vertical shaft inclined at 70°, 210 m deep and 1.8 m in diameter, was sunk from the top of the hill, and a near horizontal tunnel drilled in from the bottom of the hill, 120 m into the hill to meet the base of the shaft. Crushed stone from the top of the hill is now dropped down this vertical shaft and immediately transferred on a conveyor belt to the outside of the base of the hill where it is further treated. This 'rock by-pass scheme', as it is called, is the only one operating in the UK at the present time.

Chapter 6
Coral Reefs and the end of the Iapetus Ocean — the Silurian Period

The term Silurian System was first defined by Murchison in 1835 to cover an ancient group of fossiliferous rocks he had been studying in the Welsh Borderland. (Early workers did not distinguish between the terms system and period.) We have already discussed the great controversy which arose between Murchison and Sedgwick, and their followers, when it became apparent that the lower part of Murchison's Silurian overlapped with the upper part of Sedgwick's newly founded Cambrian. Charles Lapworth erected the Ordovician System in 1879 to essentially include the overlapping rocks and time arrival.

We now calculate that the Silurian period lasted from 435-405 million years ago, a period of fifty million years. During that time the ancient Iapetus Ocean finally closed, bringing the two halves of the British area (Figs. 33 and 39) together as one land mass. The continental collision caused the final formation of a vast fold mountain range over north west Britain, the Caledonides. This late Silurian period of mountain building is called the Caledonian orogeny. However the main earth movements over southern Britain were early Devonian and are perhaps best called Acadian. An orogeny is an episode of earth movements causing mountain building. The mountains formed are called fold mountains, e.g. the Himalayas and the Alps. The mountain range covered an area much larger than just north west Britain, since the continental masses on either side of the Iapetus Ocean included parts of North America, Greenland and Scandinavia, as well as the British area, and the Ardennes region of north west Europe.

Plate 3. The Church Stretton valley from the lower slopes of the Longmynd looking north east towards Caer Caradoc, the Lawley, and the Wrekin in the distance. The main Church Stretton Fault is about a third of the way up the side of Caer Caradoc and continues along the foot of the Lawley towards the Wrekin. Note the ridge shaped profile of Caer Caradoc in contrast to Plate 5. As only one fault occurs in the valley, downthrowing to the west, it is not a rift valley.

Plate 5. Caer Caradoc from the south end of the Lawley looking south. Note the 'volcano-like' profile in contrast to Plate 3. The Comley area, with its classic Cambrian rocks, including Comley Quarry, is immediately at the foot of the hill.

Plate 9. Steeply dipping Longmyndian rocks in New Pool Hollow with Haddon Hill in the background.

Plate 12. Cambrian/Precambrian unconformity on the Ercall with the steeply dipping basal Cambrian Wrekin Quartzite on the right resting on pink Uriconian granophyre.

Plate 13. Steeply dipping Wrekin Quartzite showing jointing and beds of various thicknesses in the top Ercall quarry. A good deal of iron staining is present. Jointing as in text Fig. 25.

Plate 14. A more distant view of the top Ercall quarry showing the boundary between the Wrekin Quartzite and the overlying green-brown Lower Comley Sandstone. The junction is two thirds of the way across the picture from left.

Plate 17. Corndon Hill from the north-east. A late Ordovician dolerite phacolith intruded into the Hope Shales along the crest of the Shelve anticline.

Plate 19. Wenlock Edge looking north east from above Craven Arms. The twin escarpments formed by the Wenlock Limestone and the Aymestry Limestone are clearly shown, the latter being in the background and also in this area forming the higher escarpment. Callow Hill with tower on right is on the Aymestry Limestone escarpment. Wenlock Edge is lower and immediately in front of Callow Hill, and in one place is in obvious shade.

Plate 21. Lenticular patch reef or 'ballstone' in Wenlock Limestone surrounded by normally bedded nodular limestones.

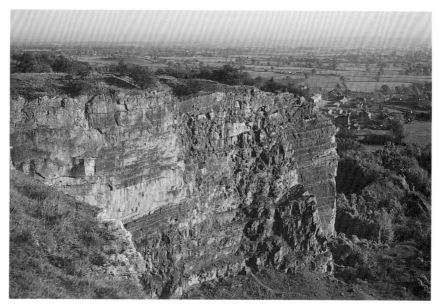

Plate 24. Llanymynech Hill showing cliffs in well bedded Carboniferous Limestone. There are two sets of obvious major joints at right angles to reach other and at right angles to the bedding, as in text Fig. 25.

Plate 26. Bridgnorth Dune Sandstone (Lower Mottled Sandstone) of Permian age showing classic cross bedding indicative of barchan sand dune deposition.

Plate 27. Upper Mottled Sandstone (Wilmslow Sandstone) at Myddle showing well spaced major joints and some cross bedding.

Plàte 28. The one remaining working quarry at Grinshill (owned by Grinshill Stone Quarries) showing the massive pale brown Grinshill Sandstone, with some cross bedding, overlain by the flaggy Waterstones and red Keuper Marl. See text for modern names of rock formations.

Plate 29. Quarry at Myddle, owned by Grinshill Stone Quarries, showing red Upper Mottled Sandstone, now being extracted again for building stone, overlain by yellow-brown, cross bedded sandstones forming the basal beds of the Grinshill Sandstone.

Plate 30. The Condover kettle hole which contained the famous mammoths. A unique photograph taken by Eve Roberts on the evening she discovered the bones in September 1986. The digger has removed peat and clay from the kettle hole and dumped it at ground level on the right. This pile of material contained the mammoth bones and had to be reworked to find them. See text and text Fig. 54 for further explanation. Steeply dipping sand and gravel forming the edge of the kettle hole can be seen to the right of the waterlogged area.

Plate 31. Current bedded fluvio-glacial sands near Buildwas at the entrance of the Ironbridge gorge.

The mountain range, once formed (Fig. 39) had a reasonably straight outline, but renewed continental drift over the last 100 million years has separated the mountain range, with various disconnected areas on either side of the Atlantic and in eastern Greenland.

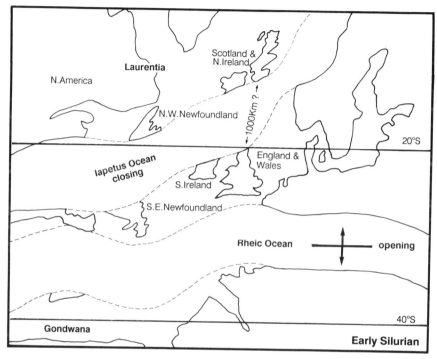

Fig. 33. Early Silurian (Llandovery) palaeogeography to show the continued closure of the Iapetus Ocean and opening of the Rheic Ocean, following on from Figs. 20 and 21. The southern Britain/south east Newfoundland microcontinent (Avalonia) has collided with Baltica and both are now moving north as one unit. Note the subtropical position of England and Wales, which in late Silurian times led to the formation of the reef-bearing Wenlock Limestone in Shropshire. The peculiar shape of north Scotland is how it appeared before movement along the Great Glen Fault in Devonian times. (After Cocks and McKerrow, 1986.)

As well as the mountain range, all the other features associated with continental collision and subduction zones were formed in the Caledonides —volcanic rocks, metamorphic rocks, deep-seated granite plutons and major fold and fault structures. All these features of this ancient mountain chain, now eroded to its very roots after 400 million years, can be studied in north west Britain, particularly in Scotland, the Lake District and Wales. Shropshire lay well within the continental margin of the southern closing land mass, and as the main orogenic effects took place near to the continental margins and subduction zones, the places or lines along which

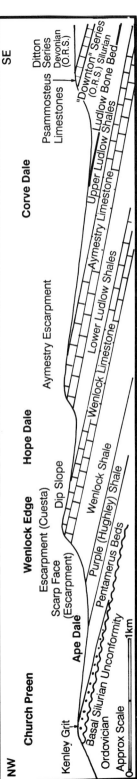

Fig. 34 Diagrammatic cross section across the Wenlock Edge area to show escarpments (Cuestas) and the classic scarp and vale scenery. Note there is no stratigraphic break (unconformity) between the Silurian and Devonian rocks, only a change of facies from marine to non-marine rocks at the level of the Ludlow Bone Bed (base of the Old Red Sandstone). The *Psammosteus* Limestones mark the base of the Devonian, but the first unconformity does not occur until the base of the Upper Old Red Sandstone (see Figs. 41 and 42 which join on to this section).

the continents joined or sutured, Shropshire only experienced slight earth movements, which led to a rise above sea level towards the end of the Silurian period. Shropshire was well to the south of the main orogenic belt and thus eventually found itself on the flanks of the newly formed mountain chain, in fact on the coastal plains at the edge of this continent, which geologists call the Old Red Sandstone continent (Fig. 39), with the Rheic Ocean to the south. No metamorphic rocks of Caledonian age occur in Shropshire, no slates, no granite masses intruded, and no major folding associated with the main, late Silurian, Caledonian orogeny, occur in Shropshire. The county simply rose up towards the end of the Silurian as part of the Old Red Sandstone continent, causing a change from marine to non-marine sedimentation, as we shall see. The late Ordovician, Shelveian (Taconian) earth movements which affected some areas, including Shropshire, on the margins of the Iapetus Ocean could be considered as an early phase of the Caledonian orogeny. However, after the Shelveian movements in Shropshire, there is no further break in the stratigraphic sequence until the base of the Upper Old Red Sandstone in the Devonian. The main 'Caledonian' movements in Shropshire are thus late Lower Devonian (Emsian), as they are in most of southern Britain, and could best be called Acadian.

During the Silurian period, before the final oceanic closure, the southern British area was steadily moving north into subtropical latitudes, and on the continental shelf area a relatively thin sequence of shallow water shelf deposits, shales and limestones was laid down, including the famous coral reef-bearing limestones of Wenlock Edge. This is the highly fossiliferous rock sequence on which Murchison founded his Silurian System, and it differs markedly from the thicker Silurian sequence formed in the oceanic and continental slope areas to the west, now found in Wales, the Lake District and south Scotland.

The Silurian period is divided into four smaller time (chronostratigraphic) divisions as follows:

'Downton' or Pridoli Series
Ludlow Series
Wenlock Series
Llandovery Series

Three of these terms are based on Shropshire rock sequences around Ludlow, Wenlock Edge and Downton Gorge. The Ludlow and Wenlock Series are accepted international time divisions of the Silurian, with their type areas in Shropshire. As the 'Downton' Series is non-marine in Shropshire it is difficult to use for international correlation, and so the Pridoli Series, based on marine rock sequences in Czechoslavakia, is the accepted international term, but the 'Downton' Series is Silurian, and not

Devonian, as was previously thought during this century in Britain. To his credit, Murchison had originally included it in his Silurian.

One puzzling feature of the Silurian in Britain is the very small amount of volcanic activity so far recognised, compared with the Ordovician, considering that subduction zones were presumably present on both sides, or at least one side, of the closing ocean. Silurian volcanics do occur in Pembrokeshire and the Mendips, but nowhere else in Wales, the Lake District and south Scotland, except for the presence of bentonite bands at many localities representing some distant source of volcanism.

The Silurian period in Shropshire

The late Ordovician Shelveian (Taconian) earth movements which affected most of Shropshire caused the sea to retreat to the west, to a north-south line from the area around Oswestry to that near Welshpool. West of this line Ashgill sediments (latest Ordovician) occur, and thus sedimentation continued through into the Silurian without a break in the Berwyns. In the early Silurian, late middle Llandovery, the sea again transgressed over Shropshire from the west to form shallow water deposits which are fairly uniform over the whole county. They have been studied in great detail and the gradual encroachment of the sea lapping around ancient shorelines can be seen clearly and will be explained later.

The best known Silurian rocks in the county belong to the shallow water shelf facies of shales and limestones laid down on the Midland Shelf Platform (Fig. 35), east of the Church Stretton Fault. Unlike the earlier Llandovery epoch, the Wenlock and Ludlow show marked differences in their rock sequences either side of the Church Stretton Fault, which had clearly become active again and marked the edge of the continental shelf. West of the fault, the Silurian sequence (except for the Llandovery) changes within 1-2 km into a thick development of graptolitic facies mudstones without limestones. These deposits are on the edge of what is called the Welsh basin, and in Shropshire can be seen in the Clun Forest (Fig. 36) and Long Mountain areas. The westward change in the Ludlow area has been particularly well studied, and shows the proximity of the continental slope to the west. It would appear that the Pontesford-Linley Fault ceased to be an active feature affecting Silurian deposition, and the Church Stretton Fault was the controlling fault line during the Wenlock and Ludlow. However, the Pontesbury-Linley Fault can be traced sporadically south west through the Clun Forest area into south Wales, where it appears to have continually disturbed sedimentation in the Silurian, as the Clun Forest Disturbance, but not affected a facies change. Woodcock (1984) has given a very detailed account of the history of the Pontesford-Linley Fault and what he calls the Pontesford Lineament,

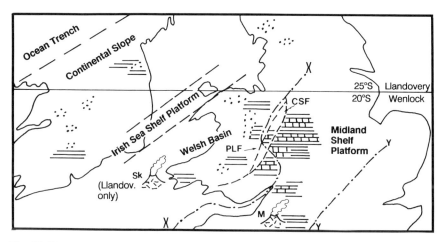

Fig. 35. Wenlock paleogeography of southern Britain. Note the formation of limestones on the Midland Shelf Platform, following its submergence by a transgression of the sea in the late Llandovery. The lines x-x and y-y mark the shoreline around the Midland Platform at the end of the middle Llandovery. M, Mendips; Sk, Skomer Volcanics of Llandovery age; ISSP, Irish Sea Shelf Platform; PLF, Pontesford-Linley Fault; CSF, Church Stretton Fault. Other symbols as in Figs. 22 and 23. Palaeolatitudes after Cocks and Fortey (1982). (After Toghill and Chell, 1984.)

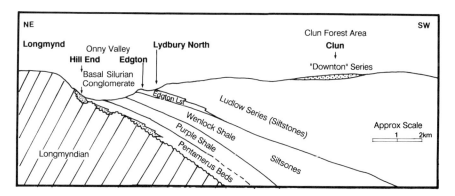

Fig. 36. Diagrammatic cross section from the Longmynd to Clun Forest to show the change in facies of Silurian sediments. Thickness of beds not to scale. (After Toghill and Chell, 1984.) Note the marked unconformity at the base of the Silurian. This is the area where Whittard mapped the early Silurian shoreline and noted old cliffs and sea stacks.

which can be traced south westward into South Wales. The 'Downton' Series also changes west of the Church Stretton Fault in the Clun Forest and Long Mountain areas, and there are no rocks at all of this age further west in Wales, where the time interval was represented by uplift prior to the main Caledonian orogeny in the Lower Devonian (Emsian).

The most typical Silurian sequence in Shropshire is that east of the Church Stretton Fault on the Midland Shelf Platform around Wenlock Edge, and here the sequence is as shown in Table 6.

Table 6. The Silurian sequence in Shropshire on the Midland Shelf Platform.

Ditton Series (Devonian) Old Red Sandstone			*Psammosteus* Limestones	
Pridoli or 'Downton' Series (Old Red Sandstone)	No Stages		Ledbury Group: Marls and cornstones	150-460 m
			Temeside Group: Temeside Shales	25-46 m
			Downton Castle Sandstone	6-15 m
			Ludlow Bone Bed	0-15 cm
Ludlow Series	Whitcliffe Beds	Ludfordian Stage	Upper Ludlow Shales	32-125 m
	Leintwardine Beds		Aymestry Limestone	25-65 m
	Bringewood Beds	Gorstian Stage		
	Elton Beds		Lower Ludlow Shales	180-260 m
Wenlock Series	Homerian Stage		Wenlock Limestone (Much Wenlock Limestone)	30-137 m
		Wenlock Shale	Coalbrookdale Formation (including Farley Member)	305 m
	Sheinwoodian Stage		Buildwas Formation	
Llandovery Series	Telychian Stage		Purple Shale	0-107 m
	Aeronian Stage		*Pentamerus* Beds	0-122 m
			Kenley Grit	0-46 m
marked unconformity				

The Llandovery transgression across Shropshire

During the middle and early upper Llandovery the sea started to transgress from the west, lapping around an irregular coastline and depositing shallow water shoreline deposits on the eroding flanks of the Longmynd and Stiperstones-Shelve area, which probably stood out as islands. The sea transgressed east past Church Stretton, finally reaching the Birmingham area towards the middle of the Upper Llandovery epoch. The varying position of this shoreline has been mapped very accurately in a classic work by Ziegler, Cocks and McKerrow (1968) following on the studies of Whittard in the 1930s. Whittard mapped the Silurian shoreline around the

Stiperstones-Longmynd area and discovered old cliff lines and sea stacks of Longmyndian rocks surrounded by Silurian beach gravels (now conglomerates) and fossiliferous shallow water sandstones around the southern Longmynd (Fig. 36).

Ziegler *et al.* went further and pioneered the science of palaeoecology, whereby the study of the size and shape of fossils, and the sediment they are found in, tells us something about the ecological conditions in which they lived, the depth of water and its temperature, the type of sea bottom and the animals' whole lifestyle. As well as this, the study of a complete fossil community or assemblage found in a rock unit tells us something about total life in the sea at the time, and is a far more interesting way of studying fossils, than simply looking at individual examples.

In their study of the Llandovery rocks, Ziegler and his colleagues found they could recognise five different animal communities characterised particularly by brachiopods, and they gave each of these five communities

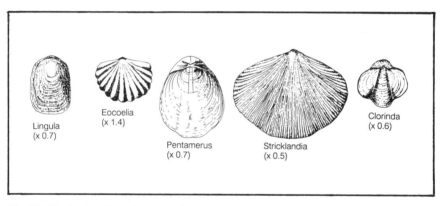

Fig. 37. The five brachiopods genera chosen to represent the five Llandovery depth communities as in Ziegler, Cocks and McKerrow (1968). Depth of water increases from left to right.

key brachiopod names, e.g. the Lingula community, but each contains a large number of different fossils. A careful study showed that each community represented a different depth of water. These five key brachiopods are shown in Fig. 37, in order of water depth. Although it is difficult to estimate absolute depths of water, it is quite easy to plot the advance of the shoreline as a deeper water community replaces a shallower one at the same locality. In this way the Llandovery transgression of the Welsh Border areas right across to Birmingham was accurately plotted.

This type of ecological study was later applied to Wenlock and Ludlow animal communities, and McKerrow's book *Ecology of Fossils* applies palaeoecology to all geological periods from Cambrian to Tertiary.

Outcrops of Silurian rocks in Shropshire

Four main areas with different rock successions can be described:

1. The Main Outcrop — the area east of the Church Stretton Fault including Wenlock Edge and Ludlow where the rock sequence is as tabled for the shelf area above.
2. West of the Church Stretton Fault around the flanks of the Longmynd and Stiperstones-Shelve area.
3. Clun Forest.
4. Long Mountain.

The Main Outcrop

This large area stretches from Ironbridge in the north-east (Fig. 4) to Ludlow in the south-west, and contains the classic ground around Wenlock Edge (Plate 19) and Ludlow. The basal Silurian rocks overlie the Ordovician of the type Caradoc area, and the highest (non-marine) Silurian beds form all of Corve Dale and pass around the south of the Clee Hills (Fig. 4) to an area near Cleobury Mortimer. Apart from local variations, the thicknesses of strata are as tabled for the shelf area. The Llandovery rocks rest with a marked unconformity on Ordovician rocks of various ages. The basal Kenley Grit is a beach gravel and shingle, now a coarse sandstone with conglomerate bands. It contains a *Lingula* community, i.e. very shallow water, and is of Upper Llandovery age, as the sea did not reach this area until that time. Its outcrop is thickest in the north, but east of Cardington it is overstepped by the *Pentamerus* Beds, and does not occur further south.

Overstep means that the sea laying down the *Pentamerus* Beds transgressed over the area in which the Kenley Grit had been deposited earlier. This feature is explained in Fig. 14. The *Pentamerus* Beds are calcareous sandstones and thin limestones crowded with the shells of the brachiopod *Pentamerus oblongus*, hence the rock name. The limestone beds are in fact fossil shell banks similar to the *alternata* Limestone in the Ordovician. They, too, are overstepped in the south, just north of the Onny Valley, by the Purple (Hughley) Shales, which have a continuous outcrop from an area north of Ironbridge to south of the Onny Valley, where they are overstepped by the Wenlock Shale. The Llandovery rocks are all shallow-water deposits indicating the slow transgression of the sea from the west, in particular by the overstepping of one formation by another as the sea encroached onto the ancient landscape.

A peculiar feature of the *Pentamerus* Beds is worth mentioning. The brachiopod is so called because a central plate, the spondylium, divides the shell into five chambers. When found fossilised, the spondylium often appears as a line dividing the angle made by the pedicle region of the shell, and produces an arrow-shaped appearance (Fig. 37). An arrow shape was commonly used in the past on government supplies and prison uniforms, and Victorian geologists christened the beds the 'Government Rock'. When breaking up pieces of *Pentamerus* Beds today, many 'arrowheads' can often be seen on one slab. If the shell has been etched out, then a mold of the arrowhead remains. This is particularly true of the *Pentamerus* Beds around Norbury, west of the Longmynd.

The *Pentamerus* Beds are followed by the slightly deeper-water Purple or Hughley shales, often containing a *Clorinda* community. By the time the Wenlock Shale was being deposited, the depth of sea was uniform over the

Plate 20. Quarry on the dip slope of the Wenlock Limestone on Wenlock Edge showing typical nodular limestones with thin shales and occasional bentonite clay bands. Good major joints planes are shown, often with iron staining.

whole of the Main Outcrop area and the outcrop is continuous from Ironbridge to Ludlow. The formation is thickest under Wenlock Edge (Ape Dale) where Basset *et al.* (1975) have defined two 'new' stages in the Wenlock Series (the Sheinwoodian and the Homerian). They have divided the Wenlock Shale into the Buildwas Formation below and the Coalbrookdale Formation above, the latter including the Farley Member (equivalent to the old term 'Tickwood Beds').

The succeeding Wenlock Limestone, renamed the Much Wenlock Limestone by Bassett *et al.* (1975), is the most famous of the Silurian rock types with its coral reef developments. It outcrops continuously from

Ironbridge in the north, where it forms the only area of Wenlock Limestone north of the River Severn at Lincoln Hill (Plate 32), to around Ludlow in the far south of the county. All the way along its outcrop it forms the conspicuous escarpment of Wenlock Edge (Plate 19). South west of Ludlow it becomes muddier and passes rapidly into mudstones of a thicker basin facies deposit. In central Wales, 50 km to the west, rocks of equivalent age are black graptolitic shales. There is a gradual passage up from the underlying Wenlock Shales, which in their upper part (Farley Member or Tickwood Beds) contain limestone nodules which become increasingly common until the base of the Wenlock Limestone is taken, where more than sixty per cent of limestone is present. The Wenlock Limestone is a nodular, irregularly bedded limestone (Plate 20) mostly made of fragmental shell material, but with some chemically precipitated limestone. It is usually thinly bedded, with narrow beds of mudstone and shale between. It is referred to as a carbonate platform limestone formed on the Midland Shelf in a warm, shallow, subtropical sea, with southern Britain at about latitude 20°S. Within the Wenlock Limestone are the famous coral reef formations now referred to as patch reefs, similar to those forming in the Caribbean today, and not a barrier or fringing reef formation, such as the Great Barrier Reef of eastern Australia.

The patch reefs (Plate 21) are lenticular, usually a fat discus shape, up to 5 m thick (sometimes thicker) and 20 m in lateral extent. The reefs are massive limestones made up of coral, sponge, bryozoan and crinoid material, and interdigitate with the surrounding nodular and bedded limestones. They are scattered about vertically within the Wenlock Limestone, increasing in size and number towards the top, but they only occur between Ironbridge and Easthope. They do not occur at all further south west, or west, where the water presumably became too deep and muddy for coral development. The presence of reefs indicates that the sea was not only warm and shallow, but also clear, since corals cannot grow in muddy seas. In the area where the reefs occur, the Wenlock Limestone is only 30 m thick, but south west of Easthope, where the reefs are absent, the Limestone thickens to 137 m.

As well as reef-building animals, brachiopods, gastropods and trilobites lived in this reef environment (Fig. 38) and are found fossilised today within the reefs, and also in the material which broke off during growth and rolled down the sides of the reef to form apron deposits.

The reefs are very pure limestone (calcium carbonate, $CaCO_3$), and are known locally as ballstones. They were exploited in the last century to provide flux for the great blast furnaces of the Ironbridge area. Nowadays, all the limestone (reef and non-reef) is quarried for a variety of uses, mostly aggregate and agricultural use. The same reefs can be found in the Wenlock Limestone around Dudley, 40 km to the east, where they are

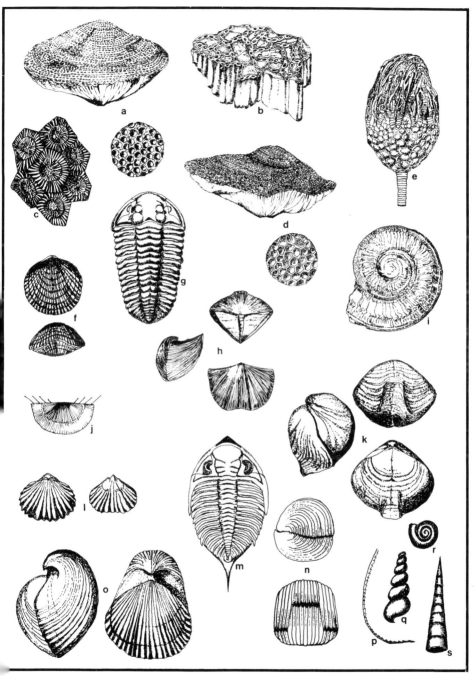

Fig. 38. Wenlock and Ludlow Series fossils from Shropshire. Corals (a-d), Crinoid (e), Brachiopods (f,h,j,k,l,n,o), Trilobites (g,m), Gastropods (i,q,r), Graptolite (p), Nautiloid (s). a. *Heliolites interstinctus*, x0.3 and x2; b. *Halysites catenularis*, x0.7; c. *Acervularia ananas*, x1.5; d. *Favosites gothlandicus, x0.3 and x2; e. Sagenocrites expansus*, x0.6; f. *Atrypa reticularis*, x0.7; g. *Calymene blumenbachi*, x0.7; h. *Cyrtia exporrecta*, x0.7; i. *Poleumita discors*, x0.7; j. *Protochonetes ludloviensis*, x1; k. *Meristina obtusa*, x0.5; l. *Microsphaeridiarhynchia (Camarotoechia) nucula*, x1; m. *Dalmanites caudatus*, x0.7; n. *Sphaerirhynchia wilsoni*, x1; o. *Kirkidium knightii*, x0.5; p. *Neodiversograptus nilssoni*, x0.5; q. *Loxonema gregaria*, x0.7; r. *Platyschisma helicites*, x0.7; s. '*Orthoceras*', x1.

called crogballs, and so this patch reef environment clearly covered quite a large area of the Midland Shelf Platform (Fig. 35). An excellent book on all aspects of limestone formation, both ancient and modern, was published by Scoffin (1987).

The succeeding Lower Ludlow Shales, Aymestry Limestone and Upper Ludlow Shales (the first of which contains a large number of thin volcanic bentonite clays) outcrop in a continuous band from Ironbridge to Ludlow. The twin escarpments of Wenlock Edge and the Aymestry Limestone, with Hope Dale between, and Ape Dale and Corve Dale on either side, are conspicuous and famous features of the Shropshire landscape (Plate 19) (Fig. 34).

An escarpment is formed by the erosion of a gently dipping succession of alternations of hard and soft layers of rock, in this case limestones and shales. The harder limestone of Wenlock Edge stands out as an escarpment, a steep 'scarp face' facing north-west, away from direction of dip, and with a gentle 'dip slope' behind sloping gently south-east parallel with the dip of the rocks, hence its name (Fig. 34). The whole feature, escarpment and dip slope, is called a cuesta, and in this area we have two cuestas formed by two limestone bands and three vales associated with the softer shales in the rock sequence. The lower part of the escarpment (or scarp face) of Wenlock Edge is actually made of soft Wenlock Shale, protected from erosion by the hard limestone capping at the top of the scarp (Fig. 34). This situation is often the case in scarp and vale scenery of this type. The term escarpment is often used to cover both escarpment and dip slope, but the term cuesta is the more correct for the whole landform.

The gentle dip slope of Wenlock Edge allows for easy quarrying of the limestone, although there is a danger of landslips as bentonite clay layers within the Limestone can be lubricated and allow large blocks of limestone to slide down-dip after quarrying has finished. On Lincoln Hill the dip is very steep — seventy to eighty degrees — and in the past the limestone was actually mined here at depth, by galleries going down the steep dip, rather than quarried. A large number of underground caverns were left behind which have led to modern subsidence around Lincoln Hill. Just to the north, the Wenlock Limestone is overlain by Coal Measures, with coal seams and ironstone nodule bands. In some shafts in this area it was possible to find all the materials needed for iron making in the one shaft — coal, ironstone, and the Wenlock Limestone underneath. The same situation occurs all around Dudley with numerous underground caverns. At the Wren's Nest, a hill north of Dudley, the galleries and chambers going down the steep dip, supported by great pillars of limestone left behind, have become unstable due to slipping along bedding planes and roof collapses. Most of those near the surface have now been filled in, or blown up and allowed to collapse. Canals were dug under the hill to meet the mines and remove the limestone. The Black Country Museum at Dudley includes a boat trip into these old caverns.

The Ludlow Series has been divided into stages (see Table 6 for Main Outcrop), and the same names are used to define the Elton Beds, Bringewood Beds, Leintwardine Beds and Whitcliffe Beds (Holland, Lawson and Walmsley, 1963) which take the place of the older terms, and are now contained in most modern works. However, the usefulness of the old rock terms (lithostratigraphic units) has been explained above, and in fact they are still used by the Geological Survey. In 1981 these stages were themselves superseded by the terms Gorstian and Ludfordian (Martinsson, Bassett and Holland, 1981). The Aymestry Limestone has no reef facies, only solitary corals, but in places contains thick shell banks of the large brachiopod *Kirkidum knightii*, (Fig. 38), particularly on View Edge.

Large disused quarries on View Edge show where the limestone was exploited in the past, and from the north eastern part of the 'edge' itself is a marvellous view north-east over Stokesay Castle along the whole length of Wenlock Edge and its associated scarp and vale scenery, as far north as Ironbridge. This view is virtually the same as shown in section in Fig. 34.

At the top of the Upper Ludlow Shales (Whitcliffe Beds) there is a change in this area from marine to non-marine facies marked by the disappearance of the exclusively marine graptolites and many brachiopods, and the appearance of bivalve molluscs, gastropods and, very importantly, the appearance of the first known British vertebrates, fossil fish, in the famous Ludlow Bone Bed. This marks the base of the Old Red Sandstone, a non-marine, fresh water, rock sequence, which continues up into the Devonian period, and was laid down in Shropshire on coastal plains fringing the continent and Caledonian Mountains formed towards the end of the Silurian period to the north-west.

The Ludlow Bone Bed is only a few centimetres thick and contains fish spines and scales. The overlying Downton Castle Sandstone, Temeside Shales and Ledbury Group of the 'Downton' Series form up to 500 m of shallow water, lake and river deposits and complete the Silurian succession.

Occasional complete fresh water fish have been found, e.g. *Cephalaspis* (Fig. 40), and the first primitive land plant remains also occur in these late Silurian rocks (Fig. 40). It is strange, perhaps, that the earliest fish should be fresh water, and not marine. What did these heavily armoured, cartilaginous, jawless fish evolve from? The sandstones and marls (calcareous clays) form the whole of Corve Dale where the red brown soils are very conspicuous. 'Downton' Series rocks also form a 'v' shaped outcrop along the axis of the Ludlow anticline between Brown Clee Hill and Titterstone Clee Hill (Figs.4 and 5), and pass round the south side of the Clees to near Cleobury Mortimer. The base of the Devonian System in Shropshire is taken at the base of a non-marine limestone sequence called the *Psammosteus* Limestones (in fact, a number of thin limestones), which forms the cap to a conspicuous escarpment overlooking Corve Dale from the east (Fig. 34).

The 'Downton' Series has, up until recently (1960s), been considered in Britain as Devonian in age because it is part of the Old Red Sandstone which in Shropshire, Herefordshire and South Wales is a non-marine facies (equivalent) of the true marine Devonian of south west England. However the Old Red Sandstone facies of non-marine continental rocks was formed as the Caledonian Mountains slowly rose up to form a new land mass to the north west, and Shropshire, lying on the southern side of this emerging land mass, clearly started to receive continental type sediments before areas further south. In Central Europe marine sedimentation continued right through from Silurian into Devonian times, whereas in the Midland Valley of Scotland, the Old Red Sandstone facies appeared in Wenlock times. Thus the correct place to draw an international boundary between the Silurian and Devonian is in Central Europe in a completely marine sequence, and not in Shropshire at the base of a diachronous rock unit, where a difference in rock type is entirely due to distant major earth movements causing a change from marine to non-marine conditions. Marine graptolite sequences in Central Europe, and elsewhere in the world, continue up into the Lower Devonian, and here the base of the Devonian period is taken at the base of the *Monograptus uniformis* Zone. In Czechoslovakia, around Pridoli, the late Silurian, early Devonian rocks are well exposed, and this is the type area for the Pridoli Series — the marine equivalent of the 'Downton' Series. On the basis of plant and fish fossils, as well as ostracods, the base of the *Monograptus uniformis* Zone in Europe probably correlates with a level near to the *Psammosteus* Limestones in Shropshire, and places the 'Downton' Series in the Silurian where Murchison originally placed it.

In the far south-east of the county, south of Cleobury Mortimer, there is a small outcrop of marine Silurian rocks near the village of Neen Sollars. The Lower Ludlow Shales, Aymestry Limestone and Upper Ludlow Shales are present, bounded by a fault on the south west. They pass up into 'Downton' Series rocks dipping north into the Cleobury Mortimer synclinal basin (Figs. 4 and 5) which has Coal Measures at its centre.

Areas west of the Church Stretton Fault System around the flanks of the Longmyndian and Shelve area
In this area (Fig. 4) the Llandovery sequence is as in the Main Outcrop, except that the *Pentamerus* Beds and Purple Shales are somewhat thicker (100 m and 180 m), perhaps indicating slightly deeper water west of the

Fault System. The basal grits have no local name and occur as lenticular masses, often quite thick. They tend to die out towards the west and cannot really be called Kenley Grit. Whittard (1932) and Ziegler et al. (1968) mapped the Llandovery in great detail and were able to show a gradual transgression of the sea from the west over a deeply eroded and irregular landscape of Precambrian and Ordovician rocks. Whittard mapped old cliff lines and sea stacks around the southern Longmynd, where coarse basal conglomerates up to 90 m thick rest with a marked unconformity on nearly vertical Longmyndian strata (Fig. 36). The basal grits pass upwards or laterally into richly fossiliferous Llandovery strata which outcrop all round the Shelve area (except in the east) and around the southern and eastern parts of the Longmynd. The *Pentamerus* Beds are particularly well exposed around Norbury, and also occur in the Shelve area at heights of up to 350 m above sea level, showing that the sea did cover all this area at some time during the Llandovery.

The succeeding Wenlock and Ludlow rocks only outcrop to the south of the Shelve area and the Longmynd, and show a marked difference from the Main Outcrop. The sequence (Fig. 36) is thicker (1,000 m plus) and, south of the Longmynd and immediately west of the Church Stretton Fault, the Wenlock Limestone is replaced by the Edgton Limestone (40-60 m) — a muddy limestone without coral reef material. The whole of the Ludlow Series is represented by 650 m of graptolitic shales and siltstones, locally with slumped beds, and with no Aymestry Limestone development. Clearly this is a deeper water sequence than the Main Outcrop, and this thick higher Silurian sequence is well exposed between Horderley and Bishop's Castle.

The strip of Silurian rocks on the southern flank of the Longmynd passes north east along the Church Stretton valley, forming the floor of the valley for all of its length, and resting unconformably on Longmyndian strata to the west. Near All Stretton, and on the west slope of Caer Caradoc (Fig. 19), there is a complete sequence in the Silurian directly west of the main Church Stretton Fault, faulted against Uriconian Volcanics. The Llandovery and Wenlock Shale are drift covered but the Wenlock Limestone, Lower Ludlow Shale, Aymestry Limestone and Upper Ludlow Shales are all exposed. The sequence here is essentially the same as in the Main Outcrop, as the area is only just west of the Church Stretton Fault System. The Wenlock Limestone contains no reef developments.

The Church Stretton Fault here must have a vertical displacement of 1,100 m to the west to bring down the Silurian strata into the Church Stretton valley. This displacement is calculated by projecting the dip of the main outcrop of the Wenlock Limestone westwards at its average dip of ten degrees. The Wenlock Limestone in this area is similar to that of the Main Outcrop directly to the south-east around Roman Bank. This would suggest that there has been no tear faulting along the Church Stretton Fault since late Silurian times. If there had been, then the Wenlock Limestone would be a different facies to that directly down dip.

Clun Forest
The Silurian of the Bishop's Castle-Horderley area merges south-west into the large basin of Clun Forest (Figs. 4 and 36). The rocks here are entirely of Ludlow Series and 'Downton' Series age (Earp, 1938, 1940) and mainly graptolitic. The Ludlow comprises a great thickness (2,000 m) of siltstones and mudstones with many slumped beds. The 'Downton' Series is represented by about 600 m of siltstones, shales and mudstones forming large outliers around Clun. The Ludlow Bone Bed and the basal beds of the Downton Castle Sandstone are represented here by a thicker sequence (6 m-12 m) of fossiliferous, silty beds known as the *Platyschisma helicites* Beds (Fig. 38) and the remainder of the sequence contains important plant remains. No beds younger than 'Downton' Series occur in the area.

Long Mountain
The Silurian sequence of Long Mountain occurs in an asymmetric syncline between the Breidden Hills and the Shelve area (Fig. 32). The main folding is post-Silurian, pre-Carboniferous, i.e. Caledonian (Acadian) although the open syncline of the Hanwood Coalfield is nearly co-axial with the Long Mountain syncline, and may have tightened the fold. Strata range in age from Llandovery right through to 'Downton' Series (Das Gupta, 1932; Palmer, 1970). The Llandovery sequence is as in the rest of the county and rests with an unconformity on Ordovician (Caradoc) rocks. The Cefn Formation (90 m) equates with the *Pentamerus* Beds, and the Buttington Formation (107 m) is similar to the Purple Shales. The remainder of the Silurian strongly resembles that of the Bishop's Castle-Clun Forest areas, being a thick, mainly graptolitic, basin-type sequence. The Wenlock is represented by the Trewern Brook Mudstone Member (457-610 m), comprising graptolitic mudstones with a lenticular development of shelly calcareous mudstones in the upper part. The Ludlow comprises up to 730 m of siltstones, often calcareous — the Long Mountain Siltstone Formation with a graptolite fauna, and with slumped beds and turbidites; and the higher Causemountain Formation comprising siltstones, which become more calcareous towards the top, with a shelly fauna. A thin bone bed marks the base of the 'Downton' Series followed by a thin development (less than 2 m) of the *Platyschisma* Limestone. Non-marine flags and coarse siltstones (180 m thick) complete the succession within the 'Downton' Series. No younger rocks occur in the Long Mountain Area.

The Caledonian (Acadian) Orogeny in Shropshire

We have mentioned earlier that a major orogeny, an episode of earth movements with mountain building, occurred at the end of the Silurian caused by the collision of continental masses on either side of the closing Iapetus Ocean. Fig. 33 shows the narrow oceanic area in early Silurian times. The Caledonian orogeny is usually considered as a late Silurian event, and indeed this is when the Iapetus Ocean finally closed. However the earth movements covered many millions of years and started in late Ordovician times in Scotland, where early fold mountains were already rising in the Grampian Highlands, and the orogeny culminated in Scotland in the late Silurian. It is in this area that we can truly use the term Caledonian, as well as in Greenland, and the rest of Laurentia on the northern side of Iapetus. However, in southern Britain, on the margin of the Avalonian micro-continent (Fig. 20), the main episode of earth movements, including folding and metamorphism in Wales and the Lake District, was at the end of the Lower Devonian (Emsian), and the orogeny in these areas and also earth movements in Shropshire are best termed Acadian rather than Caledonian. Acadian is a term used in North America for Devonian events and is now being used more frequently in Britain.

As there are no late Silurian ('Downton' Series) or Devonian rocks in the Berwyns and North Wales, we can assume that this was the period of time when the Caledonian (Acadian) orogeny folded up the area. The latest Silurian rocks are of Ludlow age, now slates, and the Berwyn Dome is thus an early Devonian (Acadian) fold structure. Caledonian (Acadian) folding also produced the Long Mountain syncline (Fig. 5) and Clun Forest fold structures, but here 'Downton' Series rocks are present, and folded, but the folding is still early Devonian. The pre-existing folds of the Shelve area and the Breiddens, in fact, all pre-existing folds west of the Church Stretton Fault, must have experienced some further folding.

The Caledonian (Acadian) orogeny had little effect in other parts of Shropshire which were even further away from the orogenic centre, particularly east of the Church Stretton Fault System. In the Ludlow area and Wenlock Edge-Corvedale area, there is a continuous passage from the base of the Silurian through to the top of the Lower Devonian Ditton Series with no unconformity present. A major unconformity (Fig. 42) then occurs at the base of the Upper Old Red Sandstone on the Clee Hills with the Middle Old Red Sandstone missing. However, the beds change to non-marine at the top of the Ludlow Series and so although no folding occurred in this area during the Caledonian (Acadian) orogeny, it obviously caused the area to rise above sea level at the end of the Ludlow as part of the Old Red Sandstone continent. The main Caledonian (Acadian) folding caused the gentle south east dip of the Wenlock Edge area and its escarpments, and fold structures such as the famous Ludlow anticline were also initiated in Middle Old Red Sandstone times and can be considered as Caledonian (Acadian). These folds were reactivated during the Variscan orogeny to affect rocks of late Carboniferous age on the Clee Hills. It is quite possible that a good deal of faulting in Shropshire is of Caledonian (Acadian) age, since many of the faults in the Wenlock Edge area do not cut the Devonian (Ditton) strata. Many of these faults could have had movement along them in the later Variscan orogeny and so their original movement is difficult to date. Movement along the Church Stretton Fault System occurred during the whole of the Silurian as it separated the western basin from the eastern shelf. Movement intensified at the height of the Caledonian (Acadian) orogeny, but the 1,100 m vertical displacement on the Church Stretton Fault referred to above was probably one of the latest movements on the fault, probably of Tertiary age. Disturbances also occurred along the Pontesford Lineament (Woodcock, 1984, 1988) in the Clun Forest area, and further south west, but no faulting occurred along the Pontesford-Linley Fault itself. A major Caledonian (Acadian) fault, the Severn Valley Fault, can be traced north east-south west, mainly under drift, from north of the Breiddens past Welshpool.

Chapter 7
All Shropshire above sea level — the Devonian period

The name Devonian System was first used by Murchison in the late 1830s when describing rocks in Devon and south west England, mainly slates but including some fossiliferous limestones, which underlay the bright red Permian sandstones of the Torbay area.

The Devonian period from 405-355 million years ago presents geologists with a major facies problem in north west Europe. Marine Devonian rocks formed on the continental shelf and slope areas of the Rheic Ocean (Fig. 39) occupy south west England, Brittany and central Europe, but in Britain, north of the Bristol Channel, the rocks change to a non-marine facies known as the Old Red Sandstone. These continental sediments formed, in Shropshire, on the coastal plains of the southern edge of the new continent formed by the continental collision following the closure of the Iapetus Ocean. This is known as the Old Red Sandstone continent because all the rocks laid down on it in Devonian times have a reddish brown colour. Further north within the Caledonian mountains, large intermontaine basins occurred in the Midland Valley of Scotland and the Moray Firth and Caithness/Orkney areas. These contained large lakes which received great thicknesses of sediments from the surrounding eroding mountains. Britain now lay almost astride the equator and the climate on this continent was very hot and arid, but with occasional flash floods.

The Old Red Sandstone is sometimes referred to as if it is a period of geological time but it is not. It is easy to fall into the trap of talking about the Lower Old Red Sandstone period, or middle Old Red Sandstone times, but it is quite correct to speak about Lower and Middle Old Red

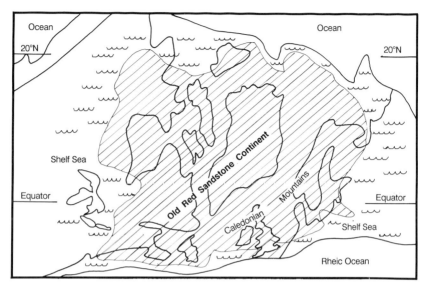

Fig. 39. Palaeogeography of the Old Red Sandstone continent in Devonian times.

Sandstone as divisions of a rock sequence. Strictly speaking the time period is the Devonian and all the Old Red Sandstone rocks are a non-marine facies, a very large rock unit, equivalent time-wise to the true marine Devonian which is best displayed in central Europe.

Comparing and correlating the rock sequence between the two facies is difficult, but many of the early fish (Fig. 40) which inhabited the fresh water of the Old Red Sandstone lakes had near relatives in the marine areas, and certain land plants, now becoming abundant, were washed out to sea and are now found in marine deposits, so correlation is possible, but not very accurate. As well as this, the shoreline on the southern fringe of the continent in the Bristol Channel area fluctuated markedly to north and south, so that in Somerset the sequence of sedimentary rocks is sometimes continental and sometimes marine, and this also helps correlation.

The Caledonian mountains over northern Britain could have been as high as the Alps or Himalayas, as they were a true fold mountain range formed by continental collisions. Their actual height we cannot calculate, but during the Devonian, when the area lay over the equator, the whole mountain range was worn away in fifty million years (by the end of the Devonian) and up to 8,000 m of coarse sandstones and flagstones were laid down in the intermontaine basins of Scotland, such as in Caithness, following the rapid erosion of the mountains. Many of the lake deposits, the Caithness Flagstones for instance, contain beautiful fossil fish, and cherts (bedded forms of pure silica) in the Grampians contain the most

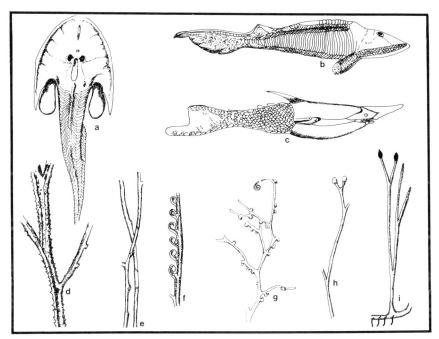

Fig. 40. Fossil fish and land plants from the late Silurian 'Downton' Series and early Devonian Ditton Series, (Lower Old Red Sandstone). a. *Cephalaspis lyelli*, x0.4; b. *Hemicyclaspis murchisoni*, x0.3; c. *Pteraspis rostrata*, x0.3; d. *Psilophyton princeps*, x0.5; e., f. *Zosterophyllum llanoveranum*, x0.7; g. *Gosslingia breconensis*, x0.4; h. *Cooksonia* sp., x0.4; i. *Rhynia* sp., x0.25, Middle Old Sandstone.

perfectly preserved fossil plants, showing the diversity of land plants during the Devonian.

South of the Midland Valley of Scotland, where some volcanoes were present, including the Cheviot area, there are no old Red Sandstone deposits preserved (except perhaps the Mell Fell Conglomerate in the Lake District), even though the area was an eroding highland area, until we reach Shropshire. Shropshire lay on the coastal plains of the continental margin with lakes, fresh and brackish water lagoons, rivers and estuaries. The Old Red Sandstone of Shropshire was all formed in this type of fresh water environment. Primitive fish and early land plants (Fig. 40) inhabited these areas, and the very earliest amphibians and land insects are of late Devonian age. The latter included flying forms, so the Devonian sees the start of the colonisation of the land and the air.

The Devonian period in Shropshire

Devonian (Old Red Sandstone) rocks in Shropshire are entirely restricted to the Clee Hills area (Plates 22 and 23). One of the obvious features of

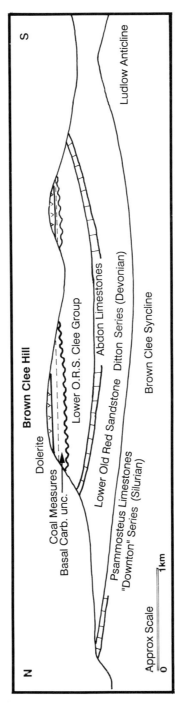

Fig. 41. Diagrammatic north-south cross section across Brown Clee Hill showing horizontal Coal Measures, with dolerite sill on top, resting with a marked unconformity on folded Devonian and late Silurian (Old Red Sandstone) rocks. This section is a continuation of Fig. 34 and also joins on to Fig. 42.

continental (or terrestrial) sedimentation is that it is erratic in its formation and patchy in its areal distribution. If you consider the Basin and Range areas of the Rocky Mountains today (Utah and Nevada), the large intermontaine basins are receiving sediment derived from the surrounding ranges, which therefore have no sediment forming on them. Thus in the Old Red Sandstone of Shropshire, in the Clee Hills area, there are breaks in the rock sequence which show that no rocks representing what we call the Middle Old Red Sandstone were formed. There is a major unconformity between the Upper and Lower Old Red Sandstone, and not

Plate 22. Brown Clee Hill above Corve Dale. *Psammosteus* Limestones escarpment is seen in the middle distance with the main mass of the hill formed by Devonian Lower Old Red Sandstone capped by horizontal Coal Measures intruded by a dolerite sill which forms the twin peaks. See text Fig. 41.

all of the Upper is present. During the time interval represented by this unconformity, the last earth movements of the Caledonian orogeny occurred, and probably initiated the fold structure of the Ludlow anticline and the gentle south east dips of the Wenlock Edge area. Fossil fish and plants allow the sequence to be relatively dated and compared with the thicker sequence of Herefordshire and South Wales, as exemplified by the great sandstone sequences of the Black Mountains and Brecon Beacons.

Now that the 'Downton' (Pridoli) Series is placed in the Silurian, the base of the Devonian in Shropshire is taken at the base of the so-called *Psammosteus* Limestones of the Ditton Series, which caps the prominent

Plate 23. Titterstone Clee Hill from above Ludlow. Like Brown Clee Hill, the hill is made of Old Red Sandstone capped by horizontal Carboniferous rocks intruded at the very top of the hill by a dolerite sill. The Carboniferous sequence is more complete than on Brown Clee. See text Fig. 42.

escarpment overlooking Corve Dale from the east. Stream sections cutting through this escarpment expose good sections of the *Psammosteus* Limestones and adjacent rocks near Morville, Monkhopton and Tugford. Above this escarpment is a large area often called the Ditton platform, from which rise the twin peaks of the Clee Hills (Brown Clee and Titterstone Clee), the highest in the county. Old Red Sandstone rocks form the majority of the high ground (Figs. 41 and 42) but the summit areas are composed of younger Carboniferous rocks, including a hard dolerite sill, which has given the hills their height.

The Devonian (Old Red Sandstone) rock sequence in the Clee Hills area is as shown in Table 7.

It is interesting to note that the fish *Psammosteus* does not occur in the so-called *Psammosteus* Limestones, but the rock name is still used. The fossil fish which gives the Limestones their name was wrongly identified originally as *Psammosteus anglicus*. It is in fact *Traquairaspis symondsi*. Of great interest in the sequence is the presence of abundant so-called cornstones and cornstone conglomerates. The former is best shown by the Abdon and *Psammosteus* Limestones but the term cornstone is often used loosely to cover both.

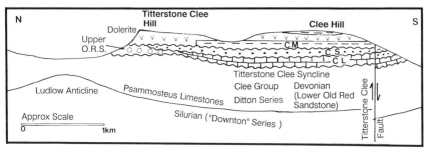

Fig. 42. Diagrammatic north-south cross section across Titterstone Clee Hill and Clee Hill showing the various folds and unconformities. Upper Old Red Sandstones rests unconformably on Lower Old Red Sandstone with Middle Old Red Sandstone missing. This is the first break (unconformity) in the stratigraphic sequence above the base of the Silurian in Shropshire, and marks the only folding of Caledonian age in the county. As it is of late Caledonian age it is best termed Acadian, and the Ludlow anticline and the Brown Clee syncline are of this age. Further earth movements and erosion caused a major unconformity to occur at the base of the Carboniferous Limestone (CL), and at the base of the Namurian Cornbrook Sandstone (CS). These were then both folded by early Variscan movements to form the shallow Titterstone Clee syncline. Further movements and erosion caused the unconformity at the base of the horizontal Coal Measures (CM), intruded by a transgressive dolerite sill of late Carboniferous age. This section follows on from Fig. 41.

Cornstones are concretionary (nodular) chemical limestones, grading from marl (calcareous clay) with many calcareous nodules, through rubbly limestones, to massive compact limestones with a high degree of purity. They range in thickness from 1 to 5 m.

Cornstone Conglomerate is made up of subrounded pebbles, up to 5 cm across, of calcareous sandstone, sandstone and concretionary cornstone, in

Table 7. The Devonian sequence in Shropshire.

Upper Old Red Sandstone	
(only present on the north side of Titterstone Clee Hill)	
Farlow Group – purple and green calcareous grits and red marls. Yellow sandstone and massive conglomerate at base	0-160 m
unconformity	
Lower Old Red Sandstone	
Clee Group – green brown coarse sandstones, a few thin marls and cornstone – conglomerates	0-400 m
Ditton Series Red and green marls with many lenticular calcareous sandstones and cornstone-conglomerates. Cornstones at top and bottom; the Abdon Limestones and *Psammosteus* Limestones respectively	600 m

a calcareous sandy matrix. They are current bedded, up to 6 m thick, cutting into the beds below. The fragments of rock often contain fossil fish. Cornstones represent fossil soils (palaeosols) similar to those found today in semi-arid areas where mean annual temperature is in the range 16-20°C, and annual rainfall is 100-500 mm. Such areas (parts of the USA, Africa, India and Israel) lie within 35° of the equator.

The Ditton Series often has a cyclical sequence as follows:

 Cornstone Conglomerate
 Cornstone
 Marl with cornstone nodules at top
 Sandstone, becoming flaggy and shaly upwards
 Cornstone Conglomerate

The effect of climate can be seen in the change from cornstone, a product of prolonged dessication, to Cornstone Conglomerate deposited by fast flowing water. The cornstones are usually mottled grey-green and purple, and the marls are red and green. The green colour sometimes occurs as small spherical patches caused by reduction of ferric to ferrous oxide which are known as 'fish eye' marls or sandstones.

The Geological Survey maps three major cornstone horizons within the *Psammosteus* Limestones covering 40 m of strata, and two in the Abdon Limestones, the lower and upper Abdon Limestone, 100 m apart.

The Clee Group contains more sandstones (green and brown) than the Ditton Series and only a few thin red marls and cornstone conglomerates. A major unconformity representing earth movements of late Caledonian age separates the Upper Old Red Sandstone from the Lower Old Red Sandstone. The Farlow Group only occurs around the village of Farlow on the north side of Titterstone Clee Hill, where its coarse basal conglomerate rests on the Ditton Series with no Clee Group present. South of the Clee Hills near to Cleobury Mortimer, the Clee Group is replaced by the Brownstones in the centre of a tightly folded synclinal basin.

Chapter 8
Shallow Seas, Deltas and Coal Swamps — the Carboniferous Period

The Carboniferous period which lasted from 355 to 290 million years ago is so named because of the abundance of carbonaceous deposits, particularly coal, formed during the latter parts of the period over many parts of the present northern hemisphere. Britain lay close to the equator at this time.

In Britain there is a well-known tripartite division into three major rock units in ascending order, the Carboniferous Limestone, Millstone Grit and the Coal Measures. These three rock units correspond roughly to the time divisions Dinantian, Namurian and Westphalian of the Carboniferous. The late Carboniferous Stephanian Series is poorly represented in Britain by rocks.

This well-known rock sequence represents, in general terms, an initial transgression of a shallow warm sea over the worn down remnants of the Caledonian Mountains, which laid down the Carboniferous Limestone (Fig. 43). This sea was later encroached on by deltas in which the Millstone Grit formed as a mainly non-marine sequence. Finally, these shifting deltas, affected by minor earth movements, supported a prolific growth of vegetation in a tropical rain forest environment. These successive growths of vegetation, covered by beds of shale and sandstone, decayed rapidly to form coal seams within the Coal Measures.

Towards the end of the Carboniferous period, uplift and a more arid climate led to the formation of late Coal Measure deposits which contain few coal seams and are essentially continental deposits. The Carboniferous sequence in south west England is totally different from that just

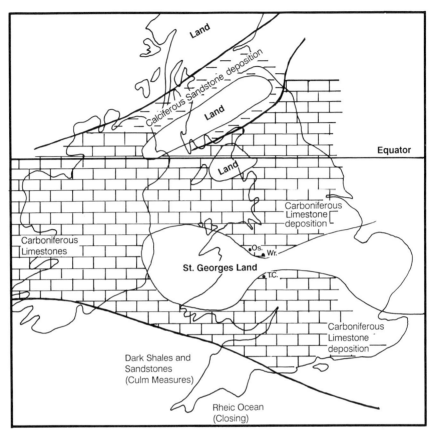

Fig. 43. Lower Carboniferous palaeogeography showing the deposition of the Carboniferous Limestone in Shropshire in two distinct areas north and south of St. George's Land. Os, Oswestry; Wr, Wrekin; TC, Titterstone Clee.

described, and is a thick (2,000 m) succession of marine mudstones and greywacke type sandstones, the Culm Measures, following the marine Devonian of the area. This sequence formed on the continental slope area of the northern margin of the Rheic Ocean (Fig. 43).

A major period of earth movements, the Variscan (or Hercynian) orogeny occurred at the very end of the Carboniferous, particularly affecting areas of south west England, south Wales, southern Ireland and western Europe, as the Rheic Ocean closed. This formed metamorphic rocks in south west England which became the northern margin of the Variscan fold mountains, into which the well-known granites of south west England were intruded. Further north, the earth movements were not so intense, although tight folds occur in south west Wales and southern Ireland, and in the English midlands only gentle Variscan folds occur, but with a large amount of important faulting. These late Carboniferous Variscan earth movements caused the British area to become a land mass once again in a now very arid, hot climate, which continued into the Permian and Triassic periods.

The Carboniferous period in Shropshire

The Carboniferous rocks of Shropshire include economic deposits of coal and iron ore, which cradled the Industrial Revolution around Ironbridge, in the Coalbrookdale coalfield. Nowadays hardly any coal, and no iron ore, is extracted from the county, but the remains of the great industrial past are preserved in the landscapes of the coalfields and the world-famous Ironbridge Gorge Museum.

At the beginning of the period, a shallow, warm tropical sea began to transgress over the remants of the worn down Old Red Sandstone continent and the Caledonian Mountains. The British area lay just about astride the equator at this time. The marine transgression was erratic to start with because the land surface was so irregular. A branch of the sea spreading from the Irish Sea area laid down Carboniferous Limestone in an area extending southwards as far as north Shropshire, whereas an area of sea over South Wales and the Bristol area (the south west province) extended just far enough northwards to deposit Carboniferous Limestone in south Shropshire. The two areas of deposition were separated by a land mass called St. George's Land which covered Wales and extended into the English midlands as the Mercian Highlands. The two areas were thus always disconnected (Fig. 43) with both limestone sequences thinning out against St. George's Land. Indeed, the two sequences are of different ages within the Dinantian Series, the rocks of the south being slightly older than those in the north, the sea having reached the south first.

Because of the proximity of land, none of the Carboniferous Limestone sequences in Shropshire are anything like as thick as in areas further north, such as the Pennines, or in the south around the Mendips, away from the shoreline.

The Carboniferous Limestone (Dinantian Series) in Shropshire

In areas north of St. George's Land on the Mercian Highlands, Carboniferous Limestone was laid down in Shropshire, and now exposed in three areas: west of Oswestry; the Lilleshall area; and around Little Wenlock, south east of the Wrekin. The first area is a large outcrop and forms the western margin of Carboniferous rocks which are the southern limit of the North Wales Coalfield. (Figs. 4 and 5). The outcrops of Lilleshall and Little Wenlock are smaller and the rock sequences thinner.

The area west of Oswestry
South west of Oswestry, Llanymynech Hill (Plate 24), with its disused limestone quarries, is the southern limit of the Carboniferous Limestone in this area. It forms a conspicuous west-facing escarpment, much quarried, with the limestone dipping away gently eastwards under younger Carboniferous rocks. This escarpment can be traced continuously northwards where, after passing into Wales near Chirk, it forms the conspicuous Eglwyseg escarpment just north east of Llangollen.

The Carboniferous Limestone succession rests with a marked unconformity on folded Ordovician and Silurian rocks of the Berwyn Dome area. The sequence can be divided into four formations. Basal Shales 30 m thick pass up into the Lower Limestone, a white rubbly limestone up to 85 m thick, well bedded and beautifully joined on Llanymynech Hill (Plate 24). The succeeding Upper Grey Limestone is 70-100 m thick, and the so-called Sandy Limestone 100 m-200 m thick completes the sequence.

The limestones are extensively quarried today around Porth-y-Waen and Llynclys Hill, 2-3 km north of Llanymynech. Although a tropical carbonate, the limestone contains no coral reefs but compound corals can be found (Fig. 44). The Upper Grey Limestone contains dark, bituminous limestone layers with very large brachiopods up to 15 cm across, *Productus giganteus*. The whole sequence is late Dinantian, Asbian and Brigantian Stages. Much of the limestone contains dolomite, calcium magnesium carbonate, as well as calcite, and is said to have been dolomitised. This is a later alteration called diagenesis, caused by the percolation of magnesium-rich liquids through the limestone after formation. The amount of dolomite varies, but never exceeds twenty per cent. It is very useful in agriculture since magnesium-bearing lime is necessary for dairy grasslands. The Wenlock Limestone of Silurian age contains no dolomite.

Limestone from Llanymynech Hill was mostly used for lime burning in the middle of the nineteenth century, 8,000 tons being produced in some years, and used for agriculture. It was burned locally but some was sent down the local canal to be burned elsewhere, e.g. at Whixall, and some quarried limestone was taken as far as ironworks in Staffordshire and used as a flux.

The Romans mined copper and lead on Llanymynech Hill, and one entrance to a Roman adit still existing today is called Ogof (cave). The veins occur within the limestone and are of late Carboniferous (Variscan) age. Chalcopyrite (fools' gold) occurs with green malachite staining, associated with galena. Zinc also occurs, and in the mid-nineteenth century lead and zinc were both mined on Llanymynech Hill, and sent by canal to

Birmingham. Zinc carbonate (calamine) was said to be more abundant than the sulphide (blende).

Lilleshall and Little Wenlock Areas
The Carboniferous Limestone of the Oswestry area dips eastward under younger Triassic rocks of the North Shropshire Plain, and probably thins out against a buried ridge of Longmyndian and Lower Palaeozoic rocks. To the east of this feature, the Carboniferous Limestone was deposited in the Lilleshall/Little Wenlock areas, which are on the western edge of the Coalbrookdale coalfield.

At Lilleshall a small outcrop of Carboniferous Limestone is 85 m thick and rests unconformably on Cambrian rocks. The sequence ranges in age from lower to upper Dinantian and so is very condensed.

Around Little Wenlock is basal sandstone, the Lydebrook Sandstone, is 25-35 m thick and passes up into a 48 m sequence of late Dinantian (Brigantian) Carboniferous Limestone. Just 4-8 m above the base of and within, the Limestone is a basal lava flow, the Little Wenlock Basalt which has a constant thickness, where it outcrops, in the area of 25 m, thickening at depth in mine shafts to 60 m in the east. The lava shows well developed columnar jointing around Doseley, and indicates a shallow submarine volcanic episode of early Carboniferous age. Similar lavas occur in the Carboniferous Limestone of Derbyshire. The columnar jointing in the basalt is similar, but not on such a grand scale, to that seen in the basalts of the Giant's Causeway in Antrim. On cooling, the lava cracks to form long columns with hexagonal (six-sided) outlines. The joints in the Little Wenlock basalt certainly show this pattern which indicates cooling in conditions of low stress. South east of the Stoneyhill Fault the basal Coal Measure unconformity cuts out the Carboniferous Limestone, the basalt and the Lydebrook Sandstone, to rest eventually on Wenlock Shales (Silurian) at the head of Coalbrookdale.

The Carboniferous Limestone succession is best studied around the Hatch area, just east of Maddocks Hill where the gently dipping Lydebrook Sandstone rests unconformably on vertical Cambrian Shineton Shales. Around here, in the limestone beds, typical Carboniferous brachiopods can be found (Fig. 44).

The Titterstone Clee Hill Area

The names of the hills in this area can be confusing. I will refer to the whole hill mass as Titterstone Clee Hill, but the southern part of the hill is also called Clee Hill, and this is where large road stone quarrying takes place today. The summit of Titterstone Clee Hill itself has the radar installations with their strange, space-age appearance. Brown Clee Hill, Shropshire's highest hill, is a separate hill mass to the north of Titterstone Clee.

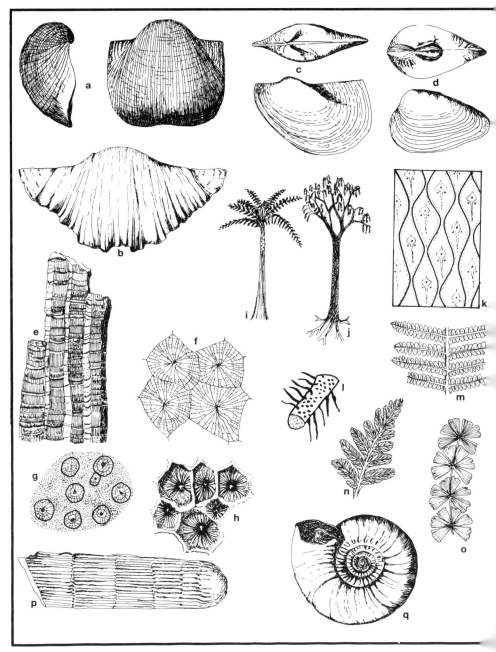

Fig. 44. Carboniferous fossils from Shropshire. Brachiopods (a,b), Non-marine bivalves (c,d), Corals (e-h), Coal Measure plants (i-p), Marine goniatite (q). a. *Dictyoclostus semireticulatus*, x0.5; b. *Gigantoproductus giganteus*, x0.3, both from the Carboniferous Limestone; c. *Anthraconaia adamsi*, x0.5; d. *Carbonicola pseudorobusta*, x0.3, both from non-marine mussel bands in the Coal Measures; e., f. *Lithostrotion vorticale*, x0.7 and x2; g. *Lithostrotion junceum*, x0.7; h. *Lonsdaleia floriformis*, x1, Carboniferous Limestone; i. Coal Measure tree fern, 8 m in height; j. Coal Measure giant lycopod *Lepidodendron*, 30 m in height; k. the same, ornament of 'trunk' or large branch, x0.5; l. *Stigmaria*, root of *Lepidodendron*, x0.1; m. *Pecopteris*, fern frond, x0.7; n. *Mariopteris*, fern frond, x0.7; o. *Sphenophyllum*, horsetail frond, x0.7; p. *Calamites*, horsetail, x0.3; q. *Gastrioceras*, typical goniatite of Coal Measure marine band, x0.7.

A narrow band of Carboniferous Limestone outcrops on both the north and south sides of Titterstone Clee (Figs. 41 and 42), the rocks being folded into a gentle syncline with a north east-south west axis. At Farlow, in the north, the Carboniferous Limestone rests unconformably on the Devonian Farlow Group. Basal Shales of Lower Dinantian age (Couceyan) pass up into 46 m of the Oreton Limestone. Grey limestones are followed by massive crinoidal limestones with some oolitic beds. Oolitic limestones are made up of thousands of tiny egg-shaped 'grains' of calcite called ooliths which are up to 3 mm across, and under the microscope have a concentric, or occasionally radial, structure. They grow on the sea bed today in shallow areas of the Bahama Banks. The most well-known British oolitic limestones are of younger Jurassic age forming the limestone escarpments of the Cotswolds. Carboniferous oolites are not so well developed, the best developed occurring in the Mendip Hills.

Brachiopods, corals and fish remains, including teeth and spines, occur in the Oreton Limestone. These fossils allow the limestone to be correlated with the Pwll-y-cwm Oolite in South Wales, of Lower Dinantian (Chadian age). The sequence is thus part of the marine south west province, south of the St. George's Land barrier.

The Namurian Series in Shropshire

The classic Millstone Grit of Namurian age follows the Carboniferous Limestone in most parts of England and Wales, away from the St. George's Land barrier, suggesting the formation of many deltas encroaching on the previously shallow clear seas. The deltas gave rise, particularly in the Pennines, to great thicknesses of current-bedded, deltaic sandstones, which are mainly non-marine, containing fresh water mussel (bivalve mollusc) bands (Fig. 44). Occasional marine incursions are indicated by thin marine bands containing fossils including goniatites (Fig. 44), which are the ancestors of the better known ammonites of the Jurassic period.

The typical coarse Millstone Grit Sandstones of the Pennines are absent from Shropshire, as the area was on the edge of St. George's Land, but Namurian sandstones do occur in the Oswestry area and around Titterstone Clee Hill, but not at Lilleshall or Little Wenlock.

Oswestry area
The Cefn-y-fedw Sandstone comprises 90 m of brown and reddish-brown sandstones, with some shales and cherts. The fauna includes, in places, abundant brachiopods and other marine fossils which allow correlation with the marine bands of the Namurian elsewhere. At Sweeney Mountain the sandstone has been extracted as a building stone and was used to restore Oswestry Church in 1807.

Titterstone Clee Hill

A thick sandstone, the Cornbrook Sandstone rests with an unconformity on Carboniferous Limestone or Devonian rocks on both the north and south sides of the hill, and is itself overlain unconformably by the Coal Measures. It comprises 200 m of coarse feldspathic sandstones and yields Namurian plant remains and pollen, but the upper part of the sequence may be Westphalian. Mid-Carboniferous (Sudetic) earth movements, an early stage of the Variscan orogeny, have caused the absence of Namurian rocks in other parts of Shropshire and the Midlands. They also reactivated the gentle north east-south west folds of the Brown Clee and Titterstone Clee synclines and the Ludlow anticline, initiated as late Caledonian structures, but reactivated to affect all rocks up to the end of the Namurian. The Shropshire Namurian is much thinner than the 2,000 m of Millstone Grit in parts of Lancashire.

The Coal Measures (Westphalian Series), Coal Measure Environments

The Coal Measures are a sequence of shales, clays and sandstones containing coal seams. Up to 3,000 m are present in some British coalfields, but the maximum thickness in Shropshire is about 750 m. Within the Coal Measures, only two to three per cent of the total thickness of strata is actual coal seams, scattered throughout the sequence. They vary in thickness, most workable seams being between 0.5 and 3 m thick, but in the South Staffordshire Coalfield the thickest coal is the Ten Yard, or Thick Coal, which is 10 m thick.

The deltas of Namurian age, which deposited the Millstone Grit of northern England and the equivalent sandstones of Shropshire, began to support a luxuriant tropical rain forest flora in Westphalian times as a humid tropical climate set in over much of north west Europe. Giant tree ferns, club mosses (Fig. 44) and other plants flourished. The rapid decay of vegetation led to the formation of swamp peats which were repeatedly covered by sediments as the deltas modified their shapes and the swamps were inundated, usually by fresh water, but sometimes marine incursions occurred. The peats were buried, compacted and turned eventually to coal. It is estimated that an eighty per cent reduction of thickness is required to change swamp peat into bituminous coal.

The cycle of swamps growing on the deltas, being inundated by muds, and then re-emerging to grow more vegetation to form more peat swamps, was repeated time and time again, giving rise to what we call Coal Measure cyclothems (Fig. 45). We find in the Coal Measures constant repetition of coal, shale, sandstone sequences followed by more coal, shale, sandstone, in that order. In the Coalbrookdale Coalfield up to thirty individual coal seams up to 2.8 m thick can be found, each indicating part of a cyclothem in a sequence of up to 200 m of Lower and Middle Coal Measures.

Underneath each coal seam is a fossil soil, called a seat earth, in which the plants grew, and it is often full of roots and rootlets. If this is a shale, it is called a fireclay and has properties, caused by the removal of minerals by plant growth, which make it economically useful for making refractory bricks. If the seat earth is a sandstone, it is called a gannister, a sandstone very rich in silica.

Above each coal seam is a waterlain shale which indicates rapid

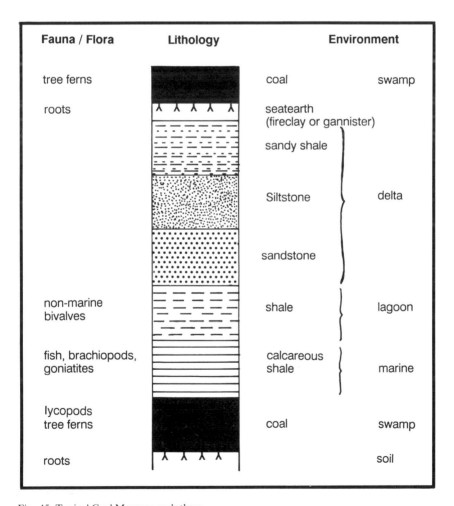

Fig. 45. Typical Coal Measure cyclothem.

subsidence of the swamp, allowing inundation. Fresh water mussels (non-marine bivalves) are quite common in these shales (Fig. 44) and can be used to correlate coal seams between different coalfields. More rarely, the inundation of the peat swamp was by a marine incursion, giving rise to the so-called marine bands in this part of the sequence, which contain goniatites (Figs. 44 and 46) and allow very accurate correlation of coal seams between the coalfields, in what is essentially a non-marine sequence.

Fourteen marine bands occur in the North Staffordshire Coalfield, and five in Shropshire. These shales above the coal seams with their non-marine, or occasional marine, bands are followed by more shales or crossbedded deltaic sandstones, which indicate a re-emergence of the land to produce deltaic sandstones, and areas for the growth of peat swamps once again. A seat earth and coal seam will follow this shale-sandstone succession and the whole cycle starts off once more.

As well as coal seams, the Coal Measures contain important deposits of iron ore, fireclays and coarse ceramic clays. The iron ore occurs as ironstones in the shales between the coal seams. The ironstones occur as nodules, variously shaped chemical concretions, concentrically banded, and formed in stagnant conditions as the swamps were being inundated. The nodules are usually siderite (iron carbonate), brown and black, and in Shropshire include the famous Pennystone Ironstone, so called because many nodules had the shape of an old eighteenth century penny. Other ironstone nodules, however, are much larger and elliptical in shape. Siderite occasionally shows a strange structure known as cone-in-cone, whereby the nodules have a columnar structure which looks like one cone sitting in another, and when split the nodules fall into conical shaped masses. Blackband ironstone is a darker carbonaceous form of siderite, and all these nodules are often called clay ironstones because they have amounts of clay and mud mixed in with them, which nowadays makes them low grade iron ores when compared with the richer sulphides and oxides occurring at other levels of the British rock sequence, particularly in the Jurassic rocks, and in rocks imported from abroad.

Fireclays, important in the making of firebricks, have been extracted from the various levels in the Coal Measures, and are still extracted today in opencast sites.

The Coal Measures are divided into three major rock units, the Lower, Middle and Upper Coal Measures. Of these, only the Lower and Middle Coal Measures contain the true cyclical Coal Measure rock sequence, with abundant coal seams, and are often called the Productive Coal Measures. After the formation of the Middle Coal Measures, widespread early Variscan earth movements caused uplift and a disappearance of most of the deltaic areas over Britain. This, together with a change to a more arid climate led to the Upper Coal Measures being mainly 'red bed' continental sediments, with only occasional thin coal seams, and they are often called the Barren Coal Measures. There is always a major unconformity at the base of the Upper Coal Measures.

The deltaic nature of the Lower and Middle Coal Measures and the erratic deposition of the more continental Upper Coal Measures, means that correlation of sequences between individual coalfields is difficult, but is helped by the occurrence of marine bands. Thicknesses of strata in each coalfield vary considerably when compared with adjacent coalfields, and it is difficult to correlate actual coal seams between coalfields, as each of these represent a different delta environment. However, it can be done with certain seams. Coal seams thicken and thin within actual coalfields, and occasionally split into two different levels, again indicating the variability of the deltaic environment.

The Shropshire Coalfields

The Coalbrookdale (Telford) Coalfield

The famous Coalbrookdale Coalfield which cradled the Industrial Revolution is divided into two halves separated by faults. The Lower and Middle (Productive) Coal Measures outcrop between two major faults trending north east-south west, and converge in the north-west near Lilleshall. The north western Boundary Fault, as it is called, is in fact a continuation of the Wrekin Fault (Figs. 4 and 5) and faults Coal Measures against Permian sandstones. On the south east side, the Lightmoor Fault bounds the Lower and Middle Coal Measure outcrop, faulting these against Upper Coal Measures. However, the Upper Coal Measure cover is thin in this eastern area and a number of mines penetrated down through them into the Lower and Middle Coal Measures. The deep gorge of the River Severn at Ironbridge, and also erosion in upper Coalbrookdale, has cut through the Upper Coal Measure cover east of the Lightmoor Fault to expose the Lower and Middle Coal Measures.

The south western margin of the coalfield is around Ironbridge and Coalbrookdale, where the Coal Measures rest unconformably on strata ranging in age from Carboniferous Limestone around Little Wenlock to Silurian Wenlock Shales in Coalbrookdale. South of the River Severn, around Broseley and Benthall, a detached area (an outlier) of Lower and Middle Coal Measures rests unconformably on Silurian rocks to the north and west, and is bounded by the Broseley and Deancorner Faults on the east and south.

Lower and Middle Coal Measures: In the coalfield the Lower Coal Measures are 20-70 m of mainly pale-coloured sandstones with subordinate grey mudstones and ironstones. The most workable coal seams and fireclays are the upper half of the sequence. In the 50-127 m of Middle

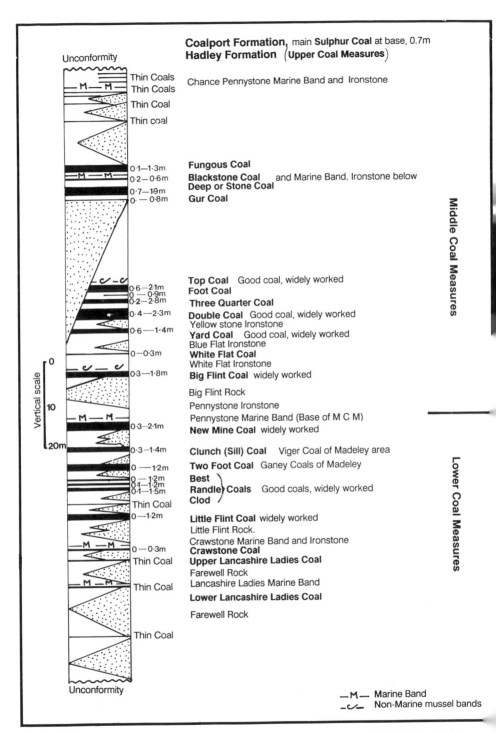

Fig. 46. Vertical section through the Lower and Middle Coal Measures of the Coalbrookdale Coalfield. All major coal seams shown with their thicknesses. Five marine bands occur in a sequence which is made up of alternations of sandstones (stippled) and shales and mudstones (blank), with ironstones and fireclays.

Coal Measures grey mudstones predominate, with many workable coal seams and ironstones occurring throughout, but with only thin sandstones and fireclays.

Numerous coal seams have been worked in the past, varying in thickness from 10 cm to 2.8 m. Twenty-one are named, many with strange local names as shown in the section in Fig. 46. These names were given to them by the miners as they worked the area. As well as these, numerous thinner seams occur. The main worked seams were the Big Flint, The Double and the Top Coals, and their thicknesses are shown in Fig. 46.

Between the seams, the succession of altenations of sandstones, mudstones, shales and fireclays contains five marine bands (Fig. 46) with goniatites, which allow accurate correlation with neighbouring coalfields, and at least two important non-marine bivalve mollusc bands with mussel shells. A number of workable fireclays and ironstone bands of siderite occur, as noted earlier. The best ironstones are the Pennystone Ironstone, up to 6 m thick below the Big Flint Coal, and associated with the Pennystone Marine Band, and another ironstone higher up, associated with the Chance Pennystone Marine Band.

The thickest and hardest sandstones, such as the Big Flint Rock (Fig. 46) have been used as local building stones, as in the case of Ironbridge Church and the Lilleshall Column. Coal Measure sandstones were also used by the Romans as the east-west portion of Watling Street passes through the area on its way westward to Uriconium (Wroxeter). The basal sandstone of the Coal Measures is usually referred to in coalfields of England and Wales particularly in south Wales, as the Farewell Rock, because it meant that the base of the Coal Measures had been reached.

The Upper Coal Measures: After the formation of the Middle Coal Measures Variscan earth movements folded the earlier strata and led to an

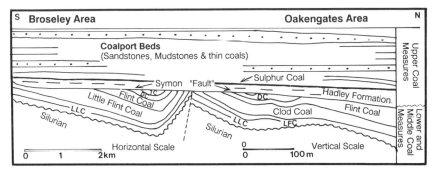

Fig. 47. Diagrammatic cross section across part of the Coalbrookdale Coalfield showing the major unconformity at the base of the Upper Coal Measures, the so-called Symon Fault. Gently dipping rocks of the Hadley Formation rest on folded Lower and Middle Coal Measures. The folding is part of the Variscan orogeny. LLC, Lancashire Ladies Coal; LFC, Little Flint Coal; DC, Double Coal; TC, Top Coal.

unconformity at the base of the Upper Coal Measures. This junction was incorrectly called the 'Symon Fault' in Midland coalfields, when it was met in many shafts. Fig. 47 shows a section across the Coalbrookdale Coalfield, illustrating the effect of these earth movements and the position of the unconformity.

A change to a more arid climate at this time caused the formation of thick continental sandstones, usually brown, reddish-brown or yellow, with marls, mudstones and occasional thin coal seams. The thickest, the Main Sulphur Coal, is 70 cm thick. Occasional thin limestones, called *Spirorbis* Limestones because of the worm which occurs in them, are also found.

This division of the Coal Measures, often called the Barren Coal Measures, is itself divided into a number of rock units, some of which occur in other parts of Britain under the same names, whereas others have entirely local names. In the Coalbrookdale Coalfield south east of the Boundary Fault, and particularly south east of the Lightmoor Fault, where the main outcrops occur (Fig. 4), the succession comprises three rock units. The basal Hadley Formation, 5-120 m thick, contains marls similar to the well-known Etruria Marl of the North Staffordshire Coalfield, and many sandstones. The overlying Coalport Formation, 95-185 m thick, comprises brown and grey-green sandstones and grey mudstones, with three thin *Spirorbis* limestones. The Main Sulphur Coal, 70 cm, is at the base of the sequence and other thin coals throughout. The Keele Beds complete the succession with 60-170 m of red, purple and brown marls and sandstone with only rare coal seams.

The Coalport Formation contains ceramic clays used for making bricks, coarse ceramics and tiles around Broseley and Coalport. Jackfield decorative tiles were particularly famous, but the even more famous Coalport China was produced using imported china clay (kaolin) from Cornwall. The china works at Coalport closed in 1926 but the china itself is still produced today at factories in the Potteries.

At Coalport, mudstones contain the famous tar deposits reached by the tar tunnel, a local tourist attraction. The liquid tar oozes out of the rocks at one horizon only in the Coalport Formation, and it is said that thousands of gallons were produced in the 1800s. It was used to coat the bottom of coracles made locally and also used in coal tar soap. A great deal was used and exported for medicinal purposes. The source of this hydrocarbon is as yet unknown, but it may originate in Carboniferous rocks at depth further north, and migrate underground to Coalport.

To the north-west of the Boundary Fault, and south west of the Wrekin, a small isolated area of Upper Coal Measures occurs around Eaton Constantine. Here the basal beds, 0-15 m thick, are called the Ruabon Marl, marls of Etruria Marl type. The overlying Coed-yr-Allt Beds comprise up to 20 m of grey mudstones and thin *Spirobis* limestones, and contain thin ironstones and a thin worked coal which may be equivalent to the Main Sulphur Coal of the Coalport Formation. The overlying Keele Beds are 45-110 m of sandstones and mudstones. The Coed-yr-Allt Beds and Keele Beds also form a faulted area between the Wrekin and Wellington.

The Coalbrookdale Coalfield supported a great number of mines in the eighteenth and nineteenth centuries, but it is beyond the scope of this book to consider any of these in detail. For further details of all aspects of the development and decline of the coalfield, the reader is referred to Barrie Trinder's book, *The Industrial Revolution in Shropshire*. Mines extracted coal, ironstone, fireclays and even limestone where they penetrated the Carboniferous Limestone or Wenlock Limestone, but their number declined in the twentieth century, and the last deep mine in Shropshire, Granville Colliery, closed in 1972. Since then, major opencast projects have taken place throughout the Coalbrookdale Coalfield area, rehabilitating the derelict mining areas, extracting coal and fireclay, and then making good the ground for the development of Telford New Town. In 1986 only one opencast site was operating, near Broseley, but others are planned for 1988-89.

The thicker coal seams and fireclays, and the better iron ores of the Staffordshire coalfields, together with the decline of the canal system and the River Severn as a navigable waterway, were some of the varied reasons why the Coalbrookdale Coalfield declined in the late nineteenth and twentieth centuries, and industry moved away. However, the atmosphere of the Industrial Revolution can still be enjoyed in the marvellous displays of the Ironbridge Gorge Museum, surely one of the world's best 'living' museums, and now a World Heritage Site.

The Oswestry Area
The Coal Measures of the Oswestry area (Figs. 4 and 5) follow on top of the sequence of Carboniferous Limestone and Cefn-y-Fedw Sandstone ('Millstone Grit') already described, and form the southern limit of the North Wales coalfield.

The Lower and Middle (Productive) Coal Measures are up to 250 m thick, mainly mudstones with thin sandstones, but with one particular sandstone up to 30 m thick, towards the base, in the south. Nine major coal seams occur up to 2 m thick. Mining in the area ceased in 1891 (New Trefonen Mine). Seven seams were formerly worked amounting to 8 m of coal in 90 m of Coal Measures at Trefonen and amounting to 3 m in 70 m of Coal Measures at Old Growern Mine. Murchison (1839) records that the two best seams were the Upper Four Foot and the Main or Six Foot Coal.

The Upper Coal Measures rest unconformably on the Middle Coal Measures and comprise three rock units. The Ruabon Marl at the base is

160 m thick comprising purple and green marls used for brick and tile manufacture. Thin *Spirorbis* limestones occur, and the sequence thins southwards. The overlying Coed-yr-Allt Beds only occur north of Oswestry, and comprise up to 70 m of sandstones and marls with *Spirorbis* limestones. The highest formation, the Erbistock Beds, equivalent to the Keele Beds elsewhere rest unconformably on the older Upper Coal Measures and comprise 200 m of purple and red sandstones with some marls, conglomerates and breccias. A few thin *Spirorbis* limestones occur, and thin unworkable coals.

The Shrewsbury-Hanwood and Leebotwood Coalfields
A large area of Upper Coal Measures (Figs. 4 and 5) occupies a belt of country between the Precambrian and Lower Palaeozoic rocks of the Longmynd-Shelve areas, and dips north under the Permian and Triassic rocks of the North Shropshire Plain. No older Carboniferous rocks are present and the Upper Coal Measures rest with a marked unconformity on the older rocks, The fault-bounded southern continuation of the Leebotwood coalfield brings Upper Coal Measures to within 3 km of Church Stretton. The Lower and Middle Coal Measures were probably never laid down in this part of Shropshire.

Two rock formations occur, the Coed-yr-Allt Beds, 120 m thick, followed by the Keele Beds, up to 450 m thick. A very thin development (up to 7 m) of Ruabon Marl occurs at the base of the sequence west of Westbury. The sequence is thus similar to that of the Oswestry area but with a major thinning of the Ruabon Marl.

The Coed-yr-Allt Beds consist of pale grey quartz sandstones, occasionally calcareous, and grey shales. In the lower half of the sequence three worked coal seams and other very thin seams, or mere films of coal, occur. The three coals are in ascending order: The Thin, or Best, 0.5 m; the Yard, 25 m higher and 0.9 m thick; and the Half-Yard, 30 m higher and up to 0.4 m thick. These have all been worked extensively in the past throughout the whole area, going back to the start of the nineteenth century. Coal from mines at Pontesford was used in smelting the lead brought down from the Shelve area close by, during the last century, and was also taken up to the lead mines themselves. Between the Yard and Half Yard coals a persistent bed of *Spirorbis* Limestone, between 1 and 2 m thick occurs. Near Pitchford the basal bed of the Coed-yr-Allt Beds is a breccia of Longmyndian Shales in a sandy matrix cemented by bitumen, hence the name given to the village. This natural bitumen has almost certainly been leached downwards from the Coal Measures, and the presence of pitch coating calcite associated with lead/barytes vein material in the Shelve area has already been referred to.

It is perhaps surprising that these coals have been so much mined in the past, but they obviously fulfilled a local need. Ironstones were also mined as well. In the Leebotwood area, mining ceased around 1880, but in the Hanwood coalfield one company went on mining successfully until 1941, working just one seam, and the Castle Peace Colliery at Pontesford did not close until 1947.

The overlying Keele Beds comprise purple and reddish-brown mudstones, marls and sandstones, occasionally calcareous. The reddish-brown sandstones are a well known local building stone and have been used extensively in Shrewsbury, for instance to build the Abbey and the Castle. The Quarry in Shrewsbury was a source of material, but could not have yielded a great deal judging by the size of the Dingle from where stone was presumably extracted. Quarries in the Keele Beds around Acton Burnell provided a sandstone for Shrewsbury buildings and also for the earlier Roman city of Uriconium at Wroxeter, the stone being transported northwards along Watling Street. It is probable that the buildings at Uriconium were 'robbed' in early medieval times to provide stone for early Shrewsbury churches, and the Abbey and Castle, as well as providing most of the stone for Wroxeter church. Blocks of Keele Beds, with the tell-tale lewis hole used by the Romans for lifting purposes, can be seen in Shrewsbury Castle and the Abbey.

Forest of Wyre Coalfield
In this extensive area (Figs. 4 and 5), encroaching into the south-east of the county, no Lower Coal Measures or older Carboniferous rocks occur, and the Middle Coal Measures rest with a marked unconformity on the Old Red Sandstone (Devonian). The Middle Coal Measures are up to 450 m thick and contain purple and green shales and sandstones, with fireclays, coal seams, and ironstones, locally called clumpers. One important marine band occurs, and a basalt sill, of late Carboniferous age, is intruded into the Middle Coal Measures near Kinlet at around the horizon of the Highley Brooch coal.

Four important coal seams occur, the thickest being the Highley Brooch which is 2 m thick. Coal mines in the area worked up until the 1950s. Just west of the south-west margin of the coalfield, near Cleobury Mortimer, is an isolated outcrop (an outlier) of horizontal Middle Coal Measures covering only half a square kilometre, and resting unconformably on Old Red Sandstone (Devonian) rocks, which are folded into a tight saucer-shaped synclinal basin. The fold structure is pre-Coal Measures in age and may be late Caledonian.

The overlying Upper Coal Measures rest with a marked unconformity on older strata. The Highley Beds (equivalent to the Coalport Beds of Coalbrookdale) are 100-200 m thick, and comprise grey shales, clays, sandstones and *Spirorbis* limestones, together with three workable coal seams, worked around the Highley area until the 1930s. The overlying Keele Beds are reddish-brown marls and sandstones up to 50 m thick. A narrow strip of Highley and Keele Beds passes northwards just west of Bridgnorth, with the Keele Beds exposed in the Bridgnorth by-pass west of the River Severn, and joins the Wyre Forest outcrop with the Upper Coal Measures of the eastern part of the Coalbrookdale Coalfield.

The Clee Hills Coalfield
The conspicuous hills of Brown Clee and Titterstone Clee (Plates 22 and 23) have their summit areas formed by horizontal Lower and Middle Coal Measures, intruded by a thick dolerite sill (Figs. 41 and 42).

On Titterstone Clee a basal Coal Measure sandstone up to 16 m thick forms a conspicuous feature on the north-east side of the hill, resting unconformably on the Cornbrook Sandstone of Namurian age, and on the west side of the hill it rests on Devonian rocks. Coal Measure shales above are up to 100 m thick, with the Four Foot and Gutter coal, 0.8 m to 1.3 m thick, lying 16 m above the basal sandstone. This seam is followed by the Smith Coal (1.2 m), Three Quarter Coal (0.6 m), and finally the Great Coal (1.5 m). Siderite ironstone modules (clumpers) are common and were exploited in the past, as were all the coal seams, usually using shallow bell pits in the eighteenth and nineteenth centuries. These are now seen as

Plate 25. A dolerite (dhustone) sill on Titterstone Clee Hill intruded into horizontal Coal Measures seen at the top of the cliff. The dolerite shows poorly developed columnar jointing.

shallow collapsed structures on the surface giving the area a cratered, lunar landscape appearance from the air. Some shafts were sunk which went through the dolerite sill to reach the Great Coal. The Coal Measures form boggy areas with rough grassland, such as around Catherton Common, a high, bleak place at times.

The Coal Measures are intruded by a late Carboniferous dolerite sill up to 90 m thick in places, but which is transgressive (Fig. 42) as it appears to rest on Lower Old Red Sandstone rocks on the north side of Titterstone Clee summit where spectacular dolerite screes occur. In this summit area the base of the sill could be parallel to the horizontal unconformity at the base of the Coal Measures which it has displaced upwards. On the south side of the hill near Cornbrook the sill appears in the Coal Measures 55 m above the Great Coal. A small patch of horizontal Coal Measures on top of the sill occurs near to Clee Hill village, but has been largely quarried away. However this upper contact of the sill can be seen (Plate 25) in the old dolerite quarries nearby, close to the main A4117 Ludlow-Kidderminster road.

The dolerite of Titterstone Clee Hill and Clee Hill — the latter name is that used for the hill area immediately north of Clee Hill village — is, when fresh, a dark blue to black olivine dolerite, a coarser grained rock than a basalt. It has always been called 'dhustone' locally, and a small village of that name can be found nearby. The word dhustone is probably a variation of ddustone, ddu being Welsh for black. Extensive quarrying took place in the nineteenth century to use the stone for stone sets, for roads and paving in particular. These were hand cut to various sizes, often 10 cm cubes. Thousands were produced and went all over the county and surrounding area. If you walk the narrow, older streets of Ludlow, you will see plenty of dhustone sets, and in Shrewsbury the station forecourt is paved with them. The rock is very hard and hardly wears at all. Clee Hill has numerous derelict quarries and on the south side of Titterstone Clee summit the huge old quarries, reaching almost to the top of the hill, show the amount of stone produced in the last century. Nowadays one large quarry near to Clee Hill village extracts the dolerite for roadstone.

In this quarry the baked bottom and top contacts of the sill with the Coal Measures can be seen (in 1986). The top of the sill is deeply weathered in places to a soft, red iron oxide-rich layer, caused by oxidation of iron rich olivine, which has to be removed before the fresh blue black dolerite can be extracted. The dolerite decays to spheroidal masses surrounded by a reddish-brown, laterite 'sandy' material, and pieces of dolerite peel off the spheres in what is called spheroidal or 'onion skin' weathering. The weathering is so deep in places that it could be due to exposure in a very arid Permian and Triassic climate, or the weathering could go back to early Tertiary times (say, fifty million years ago) when the climate was more humid and warmer than it is today.

On Brown Clee Hill two patches of horizontal Coal Measures form the summit areas resting unconformably on the Clee Group of the Lower Old Sandstone. The latter are folded into a gentle downfold, the Brown Clee Syncline (Fig. 5), whose axis trends north east-south west, and is probably a late Caledonian (Acadian) fold structure.

No Carboniferous Limestone or Namurian Cornbrook Sandstone occurs on Brown Clee, and the area may well have been one of non-deposition during the early Carboniferous.

Although not shown by the Geological Survey on their maps, a basal Coal Measure sandstone similar to that on Titterstone Clee occurs up to 16 m thick and forms a marked bench, backed by a shelf, on the north west side of the summit of Brown Clee, easily seen from around Brockton and Weston in Corvedale. About 44 m of Coal Measure clays and shales occur with four coal seams, including the Bottom Coal, 6 m above the basal sandstone, and the Foot Coal, 3 m higher up. Within a 7 m shale sequence above the Bottom Coal nodular ironstones occur, which were worked in the past and are said to weigh between a few pounds and up to half a ton. The Bottom Coal is probably equivalent to the lowest coal seen on Titterstone Clee, the Four Foot. Above the Bottom and Foot Coals on Brown Clee there are 17 m of shales and clays before three further workable seams are found; The Batty, Three Quarter and Jewstone Black, 1 to 2 m thick, separated by shales with ironstone nodules (clumpers). The coal seams and ironstones have all been worked in the past using bell pits and shallow workings on the outcrop. The small dolerite capping to Brown Clee Hill outcrops in three patches and is almost certainly the basal layer of the complete sill seen on Titterstone Clee. The dolerite on the highest point, Abdon Burf, is very weathered but forms the highest hill in Shropshire at 540 m. It is said that the hill used to be higher, but removal of dolerite by quarrying substantially lowered the height of the hill in the last century.

Coal, ironstone and dolerite (dhustone), all extracted from the summit areas, were taken down tramways north east to Ditton Priors, where a narrow gauge railway ran south east in the last century to join the main Wyre Forest GWR line (now closed) east of Cleobury Mortimer.

Carboniferous to early Permian earth movements — the Variscan Orogeny in Shropshire

As mentioned at the beginning of this chapter, a major period of earth movements occurred towards the end of the Carboniferous period affecting many parts of Europe, associated with the closure of the Rheic Ocean to the south of Britain. A large amount of folding and faulting took place over south west Britain, which lay on the northern margin of the fold mountains caused by the orogeny variously called the Variscan, Hercynian or Armorican orogeny. On a global scale this was the time when all the continental masses joined together to form one supercontinent which Alfred Wegener called Pangaea.

The Midlands and Shropshire lay just to the north of the Variscan mountain front but experienced some folding and a good deal of north-south faulting, with vertical movements giving rise to block faulting (horst and graben structures). Horsts are uplifted blocks and grabens are the rift valleys in between. This type of faulting isolated the coalfield areas and bounded them by faults, such as the Boundary and Lightmoor Faults of Coalbrookdale, and the major north-south fault which bounds the South Staffordshire coalfield on the western margin. Faults such as the Titterstone Clee Fault were probably initiated at this time (Fig. 5), and within the Shropshire Coalfields there is a large amount of faulting near to a north-south trend. Many of these faults were formed after the Coal Measures were laid down and reactivated at a later date to affect Permian and Triassic rocks and their deposition, e.g. the Pattingham-Titterstone Clee Fault (Fig. 5). One of the most important Variscan structures, just out of the area, is the Malvern-Abberley Hills fold axis (Fig. 5), a north-south line of intense folding and faulting, which also affected areas in the far south-east of Shropshire.

Many episodes of minor folding took place in the late Devonian to late Carboniferous period over Shropshire, and it is perhaps difficult to say which are Acadian (Caledonian) and which are Variscan. Within the Clee Hills (Figs. 41 and 42) the first unconformity in the rock sequence is at the base of the Upper Old Red Sandstone, and thus it would appear that fold structures such as the Brown Clee Syncline, Ludlow anticline, Titterstone Clee Syncline and the Cleobury Mortimer synclinal basin, the first of which include beds from the base of the Silurian (Llandovery) to Lower Old Red Sandstone without a break, were initiated in Devonian times (pre-Upper Old Red Sandstone) and can be called Acadian (late Caledonian).

The unconformity at the base of the Carboniferous Limestone represents further uplift at the end of the Devonian, and a further unconformity at the base of the Namurian Cornbrook Sandstone on the Clee Hills suggests early Variscan (Sudetic) movements. However, the widespread unconformity at the base of the Coal Measures, which on the Clee Hills are all horizontal resting on folded earlier Carboniferous and older rocks, suggests major reactivation of the earlier late Caledonian folds just referred to, and can be seen on Figs. 5, 41 and 42. Thus the Titterstone Clee and Brown Clee synclines were initiated in mid-Devonian times and reactivated in late Carboniferous (pre-Coal Measure) times, as Variscan movements.

The major unconformity at the base of the Coal Measures in the

Coalbrookdale Coalfield where they rest on rocks ranging in age from Carboniferous Limestone to Wenlock Shale, and also in the Wyre Forest area, suggests that in these areas a main episode of Variscan folding was pre-Lower and Middle Coal Measures. Another pulse of earth movements prior to the formation of the Upper Coal Measures resulted in a major unconformity at their base almost everywhere in Shropshire, but particularly in the Coalbrookdale Coalfield, the so-called Symon Fault. In the Coalbrookdale Coalfield the Lower and Middle Coal Measures are affected by quite tight pre-Upper Coal Measures folds (Fig. 47). Elsewhere in the county, such as on the Clee Hills, the Coal Measures are virtually unfolded. In the Oswestry area, the area furthest north west of the Malvern-Abberley line, there is little break between the Namurian Cefn-y-Fedw Sandstone and the Lower and Middle Coal Measures, and the only unconformity lies within the Upper Coal Measures at the base of the Keele Beds.

The main Variscan orogeny is usually considered in south west England to be post-Carboniferous pre-Permian, but there are few folds of this age in Shropshire, although there is an unconformity at the base of the Permian, and post-Carboniferous pre-Permian faulting almost certainly occurred. However the main Variscan orogeny did cause Shropshire and the English Midlands to be uplifted and become an arid highland area, the Mercian Highlands, which was to become even more desert-like during the succeeding Permian period.

Chapter 9
Deserts and Salt Lakes — the Permian and Triassic periods

The Permian and Triassic Systems in Britain are often taken together and called the New Red Sandstone, in contrast to the Old Red Sandstone of the earlier Devonian period. Indeed, bright red and yellow-brown sandstones are typical of these two periods, laid down in desert environments similar to those of the present day Sahara and parts of the western USA. The Permian period lasted from 290-250 million years ago and the Triassic from 250-205 million years ago.

The Permian System was named by Murchison after the town of Perm in western Russia, and the Triassic System was founded in Germany where a three-fold division of the rock sequence into Bunter, Muschelkalk and Keuper has long been established, but now superseded by more modern terms. The marine Muschelkalk in Germany (mussel limestone) is sandwiched between the non-marine Bunter and Keuper sandstone successions. The application of the terms Bunter and Keuper to the British rock sequence has caused correlation problems, particularly as the marine Muschelkalk is virtually absent, and some of our so-called Bunter is of Permian age, as will be explained later (Table 8).

During the two periods, Britain and northern Europe lay within the arid hinterland of Alfred Wegener's newly formed supercontinent Pangaea (Fig. 48) and between 20°-30°N of the equator. Consequently nearly all the sediments, and certainly most of those in Shropshire, are non-marine continental sediments with few fossils. The main marine sequence is found in north east England, where the saline Zechstein Sea laid down the Permian Magnesian Limestone which is followed by thick evaporite salt deposits and sandstone.

Elsewhere in the world the Tethys Ocean, which in the east separated the southern half of Pangaea (Gondwanaland) from the northern half (Laurasia) (Fig. 48), supported a rich marine fauna including the early ammonites in the Triassic. This marine fauna did not reach Britain until the Jurassic seas spread from the south.

The end of the Permian period closed what is called the Palaeozoic (ancient life) era and the Triassic period began the Mesozoic (middle life) era. Widespread, worldwide extinctions of many types of plants and animals, in particular marine invertebrates, such as the trilobites, occurred at the end of the Permian, to be replaced in the Triassic by new forms,

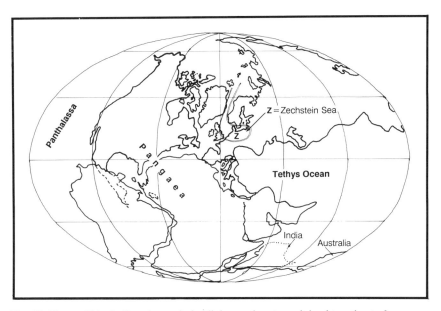

Fig. 48. The world in the Permian period. All the continents are joined together to form one enormous land mass which Alfred Wegener called Pangaea. Note the subtropical position of the British Isles. Apart from the saline Zechstein Sea there are no marine areas between North America and Eurasia and Africa. Modern coastlines are shown simply for convention. The only true oceanic crust is in the Tethys Ocean and the Pacific which together form Wegener's single ocean Panthalassa. However, Wegener did not show Tethys on his maps.

including the ammonites and abundant reptiles. These changes in the fossil record are poorly recorded in Britain because of our lack of marine deposits covering this time interval.

Many reasons have been put forward for these mass extinctions. This particular one, and another at the end of the Cretaceous period sixty-five million years ago, which led to the extinction of the dinosaurs and the end of the Mesozoic era, are the most well known.

The formation of Pangaea at the end of the Carboniferous and early

Permian led to a fall off of plate tectonic activity throughout the world, and volcanic activity and sea floor spreading at mid-ocean ridges died away, so that these ridges became less positive features on the ocean floors. This led to a widespread, well documented fall in sea level at the end of the Permian, with major regressions (retreats) of the sea from all continental shelf areas. This may have caused the widespread extinctions of shallow water marine animals in the continental shelf areas.

However, it should not have affected land plants and animals, and these two suffered extinctions, but not on such a large scale as the marine invertebrates. This theory attributing the extinctions to changes in sea level appears to be the most reasonable, and the configuration of Pangaea at this time may also have caused major climatic changes which caused extinctions on land. Another climatic change occurred slightly earlier, when a major ice age led to glaciation affecting the southern Gondwana continents at the end of the Carboniferous. The melting of these ice sheets actually raised sea level in the early Permian and caused the marine incursion of the Zechstein Sea over northern Europe in the early Permian.

Another theory to explain mass extinction at the end of the Permian, and also at the end of the Cretaceous period, suggests that a large comet asteroid (at least 10 km in diameter) or large meteorite collided with the Earth, causing catastrophic fragmentation and vaporisation of the Earth's crust in the area of impact, and hurling material into the atmosphere which cut out sunlight and produced a 'nuclear winter', but on a much greater scale. If this did happen, and sunlight was cut out for a long period of time, many forms of life might die out completely. Some animals and plants might survive so that when sunlight returned, they could evolve into new forms to take the place of those which had become extinct. This raises the question of how long the total lack of sunlight would need to last to cause these effects. Ten years, a hundred years, perhaps less, perhaps more, if a period of bombardment occurred. Certainly, the physical effects of this collision might be very difficult to detect in rocks at the end of the Permian, 235 million years ago. Rock sequences of a few metres can often represent 10,000 years or more, so the debris caused by the collision representing an instantaneous period of geological time might go undetected in the late Permian rock sequence.

Rare elements, such as Osmium, are produced by meteorite collisions and concentrations of this element have been found in rock layers associated with the late Cretaceous extinctions sixty-five million years ago, but not at the end of the Permian period.

A large impact crater should have been caused by this asteroid collision. If this had been in an oceanic area, the crater formed on the sea bed could have long since disappeared by the process of subduction. That part of the oceanic crust, along with the crater, could have disappeared down an oceanic trench, under the edge of a continent or island arc system, and passed back into the Earth's mantle. In fact if the collision did occur in an oceanic area, then it must have disappeared, as all ocean crust present today has been formed in the last 225 million years, since the break-up of Pangaea started in Triassic times. There is no pre-Triassic ocean crust present anywhere on the Earth today.

However, the collision could have been on a continental area, in which case the crater might be preserved, albeit very much eroded, or buried by later sedimentary rocks. The Manicougan crater in Quebec has been dated at around 220 million years-old and has a diameter of 70 km. This is just the size of crater which would be formed by the collision of an asteroid a few kilometres across, and so maybe this Canadian crater is the site of a catastrophic event at the end of the Permian period which caused the mass extinctions.

Other very old, large craters are now being found around the world and could explain changes in animal and plant life at other times. The Ries crater in southern Germany (Bavaria) is 24 km in diameter and dates from fifteen million years ago. Although this crater has been known for some time, it was originally considered to be the site of Miocene (fifty million years-old) volcanic activity. Recent studies have proved its impact origin with the presence of melted, shattered and stressed rocks and other impact features. In 1986 a giant crater 80 km in diameter was discovered by Chinese geologists in Inner Mongolia, 280 km north of Peking. This has been dated at about 136 million years-old, and thus coincides with the boundary between the Jurassic and Cretaceous periods when animal and plant changes occurred throughout the world.

No crater has been found to coincide with the mass extinctions at the end of the Cretaceous period sixty-five million years ago, when the dinosaurs and ammonites became extinct, and mammals then evolved rapidly to take the place of the reptiles as dominant land animals. This crater could have been in an oceanic area since subducted, and we have mentioned earlier that rock layers containing the rare element Osmium derived from meteorite impacts do occur in strata of about sixty-five million years ago. An additional theory suggests that these large impacts pierced the Earth's crust and caused the eruption of large volumes of basalt lavas from the mantle. The associated enormous increase in vulcanicity around the impact centre would add to the amount of debris and pollution in the atmosphere caused by the impact (Rampino, 1989). Present day 'hot spots' may be the sites of ancient impacts.

It should be mentioned here that the Meteor Crater, Arizona, 1.5 km in diameter is the result of a relatively recent impact which happened only 20,000 years go. It was probably not a large enough collision to produce effects outside the southern USA.

The Permian and Triassic rocks of Shropshire

The north Shropshire plain with its distinct landscape of rolling countryside and low sandstone hills is in marked contrast to the south Shropshire hill country south of the Severn. All this northern area is made of Permian and Triassic rocks forming a great synclinal basin (Figs. 4 and 5) which extends northwards into the Cheshire Plain. In north Shropshire the total thickness of these rocks exceeds 3000 m. The well known sandstone hills of Ruyton, Nesscliffe, Myddle, Grinshill and Hawkstone have yielded Shropshire's best building stones. They form three distinct lines of hills offset successively northwards by important faults, as will be explained later.

In the far north of the county large areas of late Triassic marls with thick underground salt deposits form flatter country which passes into Cheshire, but around Prees these are succeeded by the only Jurassic rocks in Shropshire. Permo-Triassic rocks also form the eastern part of the county, east of a line from Newport to Bridgnorth, a large area which could be called the east Shropshire plain.

The Permian rocks of Shropshire

At the end of the Carboniferous period the Variscan orogeny produced the Mercian Highlands over the English Midlands. Erosion of this highland area in the early Permian led to the formation towards the middle of the period of continental sandstones, conglomerates and angular scree deposits which, when harden to solid rocks, are called breccias. Two of these rock sequences are found in Shropshire. The Enville Beds (Figs. 4 and 5) which outcrop in the east of the county above the Upper Coal Measures occur in two areas. To the east of the Coalbrookdale Coalfield up to 110 m of dull red, green and brown sandstones with conglomerates and breccias rest unconformably on the Keele Beds. In the Wyre Forest area the Enville Beds vary in thickness from 100-250 m and comprise red marls, sandstones, conglomerates and breccias referred to as the Clent Breccia.

In the west of the county (Fig. 4) the second of these rock sequences occurs west of Shrewsbury around the villages of Alberbury and Cardeston and is called the Alberbury Breccia or Cardeston Stone. Here, 75 m of coarse, angular, purple and brown breccias occur with pebbly sandstones and marls. A small patch of the same rock outcrops near Pitchford. The Alberbury Breccia contains many fragments of white Carboniferous Limestone, often fossiliferous, and clearly derived from eroding Carboniferous Limestone outcrops nearby, which must have already been exposed in early Permian times around Llanymynech. Extensive and rapid erosion must have taken place at the end of the Carboniferous to strip all of the overlying Coal Measures and the Cefn-y-Fedw Sandstone in an arid

Table 8. Classification of Correlation of Permian and Triassic rocks of Shropshire showing old and new terms.

Present Classification				Old Classification		
Stages		Groups	Formations (Modern Terminology)	Formations (Old Terminology)	Old "Series"	
JURASSIC						
TRIASSIC	Rhaetian	Penarth Gp	Rhaetic	Rhaetic	Rhaetic	TRIASSIC
	Norian Karnian Ladinian	Mercia Mudstone Group	Tea Green Marls	Tea Green Marls	Keuper	
			Mercia Mudstones	Keuper Marl		
			Tarporley Siltstones	Waterstones		
	Anisian	Sherwood Sandstone Group	Helsby Sandstone	Keuper Sandstone (Ryton and Grinshill Sandstone)		
			Wilmslow or Wildmoor Sandstone	Upper Mottled Sandstone	Bunter	
	Scythian		Kidderminster Conglomerate or Chester Pebble Beds	Bunter Pebble Beds		
PERMIAN			Bridgnorth Sandstone	Lower Mottled (Dune) Sandstone		
			Alberbury Breccia Enville Beds (Clent Breccia)	Upper Coal Measures	Carboniferous	

highland area. The Alberbury Breccia is a well-known, local building stone used, for instance, to build Rowton Castle and parts of buildings as far east as Shrewsbury. Excavations on the site of Shrewsbury Abbey in 1987 show large blocks in the lowest levels of medieval buildings exposed.

All these coarse sediments overlie the Keele Beds which have late Carboniferous fossils in them. Their Permian age is deduced from the fact that they were formed after the Variscan orogeny, and in places elsewhere rocks of a similar type contain rare Permian plant and vertebrate remains. They have a patchy outcrop, probably reflecting their formation in isolated hilly areas.

This sequence is succeeded by, and passes laterally into, thick, bright red dune sandstones which blanketed many parts of the Midlands and Europe in mid-Permian times. These dune sandstones were originally considered to be of Triassic age (Table 8) and are called the Bunter Sandstone or Lower Mottled Sandstone (the latter term still used by the British

Geological Survey), but thin fossiliferous marine layers in them in north west England prove them to be of Permian age. They have now been given various local formational names throughout Britain, and in Shropshire they are called the Bridgnorth Sandstone, since they are best exposed around the town in spectacular red cliffs.

The Bridgnorth Sandstone outcrops in a semi-circular belt surrounding the southern margin of the deep Triassic-Jurassic basin of north Shropshire (Figs.4 and 5). In the east of the county, the outcrop is a north-south belt running from Newport to Bridgnorth and towards Kidderminster. The thickness is around 300 m in the north Shropshire plain area, but the formation is absent around Market Drayton, where the succeeding Kidderminster Conglomerate rests on Keele Beds, and is less than 10 m thick east of Lilleshall. From here it thickens southwards to around 200 m thick at Bridgnorth. There is a marked unconformity at the base of the Bridgnorth Sandstone which rests on rocks ranging in age from Cambrian to Upper Coal Measures or Enville Beds/Alberbury Breccia.

One has to imagine areas of sand dunes (dune fields) spread over a deeply eroded, rock landscape about 260 million years ago. The fossilised dune sandstones have the structure of barchan dunes, which can be seen today in the Sahara and other sandy deserts. They have a horseshoe shape in plan view and migrate across the desert, blown by a persistent prevailing wind. Within a barchan dune the bedding is referred to as cross bedding, as the sand is blown up the steep windward side of the dune and down the gently sloping leeward side.

The migrating dunes cut across each other and cover older dunes, so that a sequence of cross-bedded sandstones (Fig. 49) is built up in a modern dune field. In the Bridgnorth Sandstone the sequence of cross-bedded sandstones is beautifully shown in the cliffs around Bridgnorth (Plate 26),

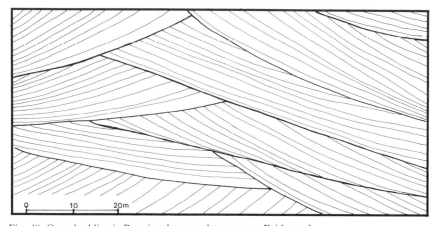

Fig. 49. Cross bedding in Permian dune sandstones as at Bridgnorth.

and one can imagine that the area was once just like the Sahara desert is today. Studies of the dune structures show that the prevailing wind at the time (palaeowind) was from the east during the mid-Permian over the whole of Britain, from Exeter to Elgin on the Moray Firth.

The sandstone is made up of spherical grains of quartz 1-2 mm in diameter, which under a hand lens can be seen to be beautifully rounded and frosted (as in frosted glass) by wind action which allowed each grain to rub against another. These wind blown, or aeolian, sandstones are sometimes called millet seed sandstones. Each grain of quartz has been coated with red iron oxide (haematite) which in solution percolated through the sandstone at a later date, and also formed a cement to harden the rock.

However, the rock is quite soft and can be rubbed away with the finger, although it has been used in the past to build many churches and other buildings around Bridgnorth, and to the east and south. Many of these buildings are now themselves suffering from erosion, including wind erosion!

The Triassic rocks of Shropshire

Further uplift of the existing desert highlands in the early Triassic led to a major river system flowing northwards from Brittany into southern England and into the Midlands. This major river and its tributaries laid down a series of coarse river gravels, now conglomerates. These deposits, the Bunter Pebble Beds of earlier authors, are of early Triassic age and were formed as river channels cut into the older dune sandstones. East of Market Drayton the Pebble Beds rest directly on Keele Beds. In Shropshire these beds are now called the Kidderminster Conglomerate and, further north, the Chester Pebble Beds. They follow the outcrop pattern (Fig. 4) of the underlying dune sandstones and vary in thickness from 130 m around Bridgnorth, to 200 m east of Oswestry and 100 m around Wem, but only 45 m north of Shrewsbury.

In modern terminology (Table 8) they are the basal unit of the Sherwood Sandstone Group (Warrington et al., 1980), the old term Bunter no longer being used.

The pebbles are mainly pale coloured quartz and quartzite, well rounded, in a brown red sandstone matrix. Some of the quartzites contain Ordovician fossils and have been derived from as far away as northern Brittany, where outcrops of Ordovician quartzites with abundant fossils occur. Pebbles of more local rocks also occur. The elongated pebbles are often up to 10 cm along their longest diameter, and around Bridgnorth the old river channels in which they were formed can be seen, cutting down into the dune sandstones, east of the river at the top of the Hermitage Hill on the 'old' Wolverhampton road. Near Hinstock, in the north of the county, a remarkable find of fossil silicified wood was made in the early 1900s. Logs 15-22 cm in diameter and up to one metre in length were found embedded in pebble beds and almost certainly derived from local Upper Coal Measures.

The Kidderminster Conglomerate is followed by the Wilmslow or Wildmoor Sandstone Formation (the Bunter Upper Mottled Sandstone of earlier authors). This formation contains 100-200 m of bright and deep red cross-bedded sandstones, probably laid down in a mixture of braided river and dune environments. Large quantities of this red sandstone have been extracted for use as building stone around Harmer Hill, Webscott, Myddle (Plate 27) and Nesscliffe. In recent years, Grinshill Stone Quarries have reopened a quarry at Myddle (see later).

The overlying Helsby Sandstone and Tarporley Siltstone Formations approximate to the old Lower Keuper Sandstone. The term Keuper should no longer be used in the division of the British Triassic, but it still survives in describing late Triassic rocks in particular. These formations vary in thickness from 120 m in the east, 40-60 m around Nesscliffe, to about 25 m around Grinshill, and about 60 m around Hawkstone. In the east and south-east of the county the sequence is mainly red and brown sandstones with flaggy Basement Beds separating the massive sandstones from the underlying Upper Mottled Sandstone. In the north Shropshire Plain, on the sandstone hills, the sequence has been more precisely divided into the lower Building Stones division of up to 45 m of sandstones, the famous Grinshill and Ryton Sandstones (Helsby Sandstone Formation), overlain by 3-16 m of the 'Keuper' Waterstones (Tarporley Silstone Formation).

The Ryton and Grinshill Sandstones are the best known building stones in the area, and comprise yellow and creamy-brown, massive sandstones in the Grinshill area which become redder and less hard further west around Nesscliffe. The conspicuous pale colour around Grinshill and Hawkstone has probably been caused by Tertiary igneous activity about fifty million years ago, which not only formed the dolerite dykes and mineralisation of the area (see later), but also chemically altered the oxides in the sandstones from red ferric to paler ferrous oxides and hardened the rocks. The sandstones have widely spaced major joints, few bedding planes, and are even grained (Plate 28). They can thus be extracted in large rectangular blocks and easily dressed and cut, and are called freestones by stone masons. There is some cross bedding present, and it is still debated whether the beds are water-lain or wind-blown deposits. Rare, well preserved bones of the reptile *Rhynchosaurus* have been found and are now kept in Shrewsbury Museum at Rowley's House.

Very large, disused quarries show where enormous quantities of pale creamy-brown sandstones were extracted in the last century at Grinshill.

However, around Myddle, Nesscliffe and Ruyton the sandstones are redder and quarries occur in the dull red freestones of this horizon, and the underlying Wilmslow Sandstone (Upper Mottled Sandstone). Only small quantities of pale building stone were taken from the Hawkstone area. Red stone from Nesscliffe has been used less widely away from its local area, but has been used for later restorations of churches such as St. Mary's in Shrewsbury, the Abbey and Wroxeter church. The red sandstones at the foot of Grinshill have been used as building stones and can be confused with those from Nesscliffe and Myddle.

The pale, creamy-brown sandstone from Grinshill was used extensively throughout the county and beyond. In Shrewsbury the English and Welsh Bridges are built out of it, as well as the Old Market Hall in the Square, the Column, the Library (the old Shrewsbury School) and the station, to name just a few of the buildings, and of course many churches are built of Grinshill stone either completely or in part. Parts of Chequers, the Prime Minister's country residence, and also Downing Street, are also built of Grinshill Stone.

It is worth listing here the detailed section of sandstones in the main Grinshill Quarry as measured by the Geological Survey in 1920:

		Feet	Inches
Keuper Marl Red marl, rubbly, with thin sandstone-bands, the lower 6ft. more sandy	seen to	10	0
Waterstones			
Flag Rock: sandstone, light yellow and greyish, fine fairly evenly bedded and weathering into flags, formerly used for roofing		20	9
Esk Bed: yellowish sand with black specks of manganese dioxide; harder in patches and forming a speckled sandstone			9
Grinshill Sandstone			
Hard Burr: sandstone, grey, fairly hard, fine grained, with black specks of manganese dioxide; used for doorsteps, paving, curbs, etc; top even, base uneven, thickness variable, up to		3	0
Building Stone: sandstone, pale yellow, fairly hard, fine grained, with irregular joints ('shakes' and 'cricks'). This bed yields the best stone		15	0
Building Stone: sandstone, pale yellow to greyish, rather soft, fine grained, with irregular joints		10	0
Building Stone: sandstone, yellow, coarse grained, rather too soft and porous, with irregular joints		12	0
Sandstone, yellow, soft, with patches of reddish sand that make it unsuitable for building	seen to	20	0

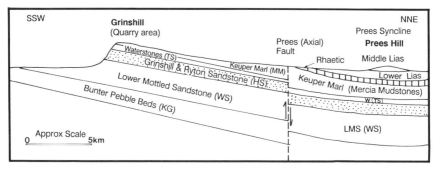

Fig. 50. Diagrammatic cross section from Grinshill to Prees showing the distribution of Triassic and Jurassic rocks.

One quarry at Grinshill now remains in production (Plate 28) and exposes the top 9 m of Grinshill Sandstone, showing marked cross bedding, the full 7 m thickness of the flaggy Waterstones and the basal 3 m of the red Keuper Marl. In recent years Grinshill Stone Quarries have expanded their production at this quarry and provided building stone for restoration work on numerous buildings in Shropshire and elsewhere.

In Shrewsbury the restoration of the Library shows the pale Grinshill Stone in all its former glory, as does the work carried out in the restoration of Rowley's Mansion. Like all building stones, the Grinshill stone is affected by weathering and pollution. However it is the most durable of stones and, although it does become darker after many years of exposure, cleaning quickly restores the classic pale colour. Grinshill Stone has also been used recently in Shrewsbury at: St. Chads; the Castle; the Music Hall; the new Trustee Savings Bank building in Swan Hill. The stone has also been used for fine internal decoration in Shrewsbury's Catholic Cathedral.

Further afield, Grinshill Stone Quarries are sending stone to: Worcester and Hereford cathedrals; Warwick Castle; St. Chads, Birmingham; Chatsworth House; Chester Castle and Magistrates' Court; York Museum and Botanical Gardens; Biddulph Gardens, Stoke.

Not all the stone used, both locally and elsewhere, is the classic pale brown sandstone. Grinshill Stone Quarries have recently reopened a quarry at Myddle in the red Upper Mottled Sandstone, and from here they obtain good quality red building stone, equally as hard as that at Grinshill. This has been used at many of the above localities, if red stone has been required. This quarry, 1 km east of Myddle, on the sandstone escarpment stretching to Webscott and Harmer Hill, is of interest as it exposes the boundary between the red Upper Mottled Sandstone and basal yellow brown sandstones of the Grinshill and Ryton Sandstones. In the western corner of the quarry, 6 m of massive dull red sandstones with very few bedding planes, and a few major joints (Plate 29), the source of the good red stone, passes abruptly up into 2 m of thinly bedded, and quite obviously, cross-bedded pale yellow, brown sandstones. The colour change does not exactly coincide with the marked change to cross-bedded sandstones, as the red colour does pass up into some of these overlying beds, the colour boundary being an irregular line. These basal Grinshill and Ryton Sandstones are of no use for building and presumably pass up into the main building stones seen at Grinshill.

Separating the Grinshll Sandstone from the overlying Waterstones (Tarporley Siltstone Formation) is a conspicuous thin bed, called the 'Esk' Bed, only seen east of Baschurch and in the Grinshill area. It comprises only 15-45 cm of yellow and grey sands and sandstones with an ashy appearance, speckled with black Manganese dioxide and sometimes with a barytes cement.

The overlying Waterstones (Tarporley Siltstone Formation), seen to good effect on Grinshill (Plate 28) (Fig. 50) are more thinly bedded grey, yellow and brown, occasionally pink, fine-grained, micaceous sandstones and flagstones with very thin green micaceous siltstones. They show irregular bedding and sedimentary structures, including ripple marks, suggesting deposition in lakes. In other parts of the Midlands rare *Lingula* brachiopods suggest an inter-tidal environment. In the North Sea area and Europe, beds of this age are truly marine and represent the Muschelkalk of

Fig. 51. Triassic reptiles from Grinshill. a. *Chirotherium* and footprints. No bones of this animal had been found until 1965 when some foot bones of this track maker were discovered in Germany. Length about one metre. b. *Rhynchosaurus*. Two well preserved skeletons are kept in Shrewsbury Museum as well as some footprints. Length about 0.35 metres.

the old German tripartite division of the Triassic into, Bunter, Muschelkalk (Mussel Limestone) and Keuper. The proximity of the Muschelkalk sea raised humidities so that horsetails and conifers grew in the desert areas of the east Midlands and North Sea. Land reptiles inhabited the areas, including early herbivorous dinosaurs. In the Waterstones at Grinshill *Rhynchosaurus* bones and footprints have been found, the latter formed in soft muds at lake margins. Other footprints also occur, which until 1965 could not be matched with any known skeleton, and these are trace fossils of an animal whose shape could only be guessed at. From the shape of the prints the animal has been named *Chirotherium* (hand beast) and a reconstruction of it is shown in Fig. 51. However, in 1965, bones of this track maker, a partially bipedal pseudosuchian reptile were found in Germany. Rain pits occur in the Waterstones, and cavities with crystal outlines show where rock salt (halite, sodium chloride) and anhydrite (calcium sulphate) crystals once occurred, so the lakes were obviously saline.

The Waterstones are probably so named because of the glint of light on the white mica- (muscovite) covered bedding planes which resembles 'watered silk'. The rock is not so named because it is an aquifer. There must have been a large mass of granite or schist being eroded away nearby, rich in muscovite, to provide the source for these micaceous flagstone.

The flaggy Waterstones at Grinshill are today quarried for ornamental flagstones and walling stones, and some blocks of poorer building stone are sold for rockery stone.

The Grinshill and Ryton Sandstones, and the Waterstones, form the conspicious low sandstone hills of the North Shropshire Plain such as Grinshill, Hawkstone, Pim Hill and Nesscliffe. The outcrop pattern is affected by three major faults (Figs. 5 and 4) which displace the outcrop, and the lines of hills, successively northwards. These are: the axial or Prees Fault, which follows the axis of the North Shropshire Triassic synclinal basin; the Brockhurst Fault; and the Hodnet Fault. These are all probably of post-Jurassic age as one of them, the Prees Fault, cuts Jurassic strata. However, they could be older faults affecting the Palaeozoic rocks further south, since reactivated in Tertiary times. The Ruyton to Nescliffe ridge is displaced north by the axial Fault to form the Myddle to Grinshill ridge, and this in turn is displaced north by the Brockhurst Fault to form the Hawkstone to Hodnet ridge.

Because the rocks dip gently northwards into the North Shropshire syncline, the hills form conspicuous escarpments facing south or south west (Fig. 50). The panoramic view from the top of Grinshill (or is it Grinshill Hill — maps do not agree on this) southwards across the older Permian and Triassic rocks and towards the Lower Palaeozoic rocks of the South Shropshire hills is quite magnificent.

It is very difficult in the west of Shropshire around Nesscliffe to distinguish the Wilmslow Sandstone (Upper Mottled Sandstone) from the overlying Ryton and Grinshill Sandstones (Lower Keuper Sandstone). Both groups yield good building stone but have a similar colour in this area as seen on Nesscliffe Hill. Here a 30 m high quarry face shows no abrupt change and yet while the lower beds are hard enough to be quarried for building stone, they are deep red in colour and clearly part of the Upper Mottled Sandstone. They pass upwards into slightly harder and less bright red sandstones referred to as Ryton and Grinshill Sandstones.

However, between Myddle and Webscott the Upper Mottled Sandstone, which has been extensively quarried, is a red freestone and is followed, at Myddle, in the quarry recently reopened by Grinshill Stone Quarries (see earlier), by yellow sandstones of the Ryton and Grinshill Sandstones, and so here a more obvious colour change occurs. Between Webscott and Harmer Hill a number of 'cliff' houses have been built up against old quarry faces in the red Upper Mottled Sandstone.

This colour change becomes more marked around Grinshill and Hawkstone. The Grinshill sequence has already been described, but that around Hawkstone, east of the Brockhurst Fault, deserves description. The Upper Mottled Sandstone (Wilmslow Sandstone), comprising red and yellow mottled sandstones, covers a large area around Weston-under-Redcastle and Wixhill, and including the Citadel and Bury Walls. This area is bounded to the south by the striking escarpment of the Upper Mottled Sandstone from Weston Heath Coppice east to Hopley Coppice. Overlying these red sandstones 40 m of white and pale yellow sandstones occur which can be assigned to the Ryton and Grinshill Sandstones, and can be seen capping a second escarpment formed by mainly red sandstones at The Red Castle, and also at Grotto Hill. At Grotto Hill the 45 m of pale sandstones is the thickest development of the Ryton and Grinshill Sandstones in Shropshire, with much disseminated barytes, and overlying bright red Upper Mottled Sandstone (Wilmslow Sandstone). These pale-coloured sandstones rapidly thin eastwards towards Marchamley Hill where they are only 6 m thick, overlain by the Esk Bed of the Grinshill sequence and the Waterstones. The Waterstones comprise red and grey flags and thinly bedded sandstones which are 16 m thick, twice the thickness seen at Grinshill. They cover a large area around Marchamley and south-eastwards. Further east still the pale sandstones disappear, and at Kenstone Hill, on the steep escarpment, the Waterstones directly overlie bright red Upper Mottled Sandstone and the Ryton and Grinshill Sandstones have disappeared. The sequence along the escarpment now reaches the Hodnet Fault where it is cut out, and around Hodnet and further east all exposures are in the older Bridgnorth (Dune) Sandstone and the Kidderminster Conglomerate (Bunter Pebble Beds).

Thus in the Hawkstone-Weston area we have two escarpments, one to the south formed by the Upper Mottled Sandstone and a second further north formed by Ryton and Grinshill Sandstones overlying the Upper Mottled Sandstone. The 'plateau' area in between the two escarpments is occupied by the villages of Weston and Wixhill. This is a different situation to that seen at Grinshill and Nesscliffe, and also in this area we see the rapid eastward thinning of the Ryton and Grinshill Sandstones and a corresponding marked thickening of the Waterstones.

The Waterstones are succeeded by a great thickness, up to 1,400 m, of mudstones, the Keuper Marl, now renamed the Mercia Mudstone Group, of which the Waterstones are actually the lowest formation. The Keuper Marl (Fig. 50) underlies extensive areas of north Shropshire completing the very thick sequence of Permo-Triassic (New Red Sandstone) rocks of the north Shropshire Plain. The top of the Triassic sequence can be seen in north Shropshire where marine Jurassic rocks follow on, but in the east of the county, only small areas of Keuper Marl occur, east of Telford, and the top of the sequence is nowhere seen in this part of the Midlands into Staffordshire.

A return to more arid conditions and a retreat of the quasi-marine conditions which formed the Waterstones led to formation of playa lakes and other saline lakes, in which great thicknesses of salt deposits, particularly rock salt, formed in North Shropshire and Cheshire. Playa lakes are very shallow sheets of water which appear on low ground after sudden and violent rain storms in deserts, such as in the western USA today. They may cover hundreds of square kilometres and yet only be a few centimetres deep. They repeatedly dry out and leave mud flats rich in salts, like the Bonneville Salt Flats in the USA. While still wet, the muds record rain pits and tracks of desert animals, as well as sun cracks. Playa lakes and deeper saline lakes give rise to halite (rock salt), gypsum and anhydrite, and are also valuable sources of nitrates and borax.

In North Shropshire the Mercia Mudstones are about 1,000 m thick around Wem, increasing to 1,400 m thick around Whitchurch, but this thickness includes two separate salt bearing (saliferous) horizons, 200 m and 400 m thick. In the east of the county only the lowest beds of the sequence are seen and no distinct salt beds occur.

The 'Keuper' Marl comprises mainly reddish brown 'chocolate' coloured mudstones, and silty mudstones with thin sandstones called 'skerries'. Although traditionally called marls, which suggests a high calcaerous (lime) content, the rocks are only slightly calcareous, apart from the highest Tea Green Marls, and the term Mercia Mudstone Formation is a better one. The sediments are mainly clay minerals mixed with wind blown (aeolian) quartz dust and crystals of dolomite and gypsum. Green blotches within the red marls are referred to as 'fish eyes', caused by spherical zones of decolourisation around a black central core, probably a radioactive or diagenetic mineral.

As well as fish eye marls, grey, buff and pink sandstones occur called skerries, consisting of quartz, some feldspar and worn evaporite salt crystals. The skerries contain ripple marks, sun cracks on bedding planes, salt pseudomorphs and rain pits, all indicating great aridity, with occasional flash floods causing the playa lakes to form.

There are very few natural exposures of the Keuper Marl in the north of the county, where the cover of glacial deposits (drift) of the last Ice Age and other superficial deposits is very thick, often up to 60 m. Around Whitchurch, and north into Cheshire, the sequence within the Keuper Marl, entirely proved by exploratory boreholes for salt is as follows:

Upper Keuper Marl:	red, green and grey mudstones and much anhydrite	160 m
Upper Keuper Saliferous Beds:	red, green and grey mudstones with thick halite deposits and a little gypsum	400 m
Middle Keuper Marl:	marls with a little anhydrite	350 m
Lower Keuper Saliferous Beds:	marls with thick halite	200 m
Lower Keuper Marl:	marls with some anhydrite	270-320 m

In a temperate, wet climate salt beds can never occur (outcrop) at the surface and probably do not anywhere reach the base of the superficial drift because of solution by downward percolating water. They are present at depth against a surface which represents the lower limit of groundwater circulation. This 'outcrop' is overlain by a zone of fragmented (brecciated) mudstone derived partly from the collapse of marls above the saliferous beds, and from inter-bedded mudstones within the saliferous beds.

Whitchurch itself sits on top of the Upper Keuper Saliferous Beds, but the solution of salt at depth has already occurred and so the situation is relatively stable. Further north, extraction of salt led to a good deal of subsidence but this problem has nowadays been cured. The upper part of the Keuper Marl (Mercia Mudstone Formation) is coloured green and grey and called the Tea Green Marls, a rock formation which can be traced from south west England to Yorkshire, and yet is only a few metres thick, 15-18 m in north Shropshire. This formation is nowhere exposed on the surface in north Shropshire, being covered by drift, but has been proved in boreholes around the Jurassic area of Prees and Audlem. The colour change occurs over the whole of Britain at this horizon, and the green and grey mudstones are probably of marine origin.

The colour reflects a higher initial organic content than the playa lake deposits below, promoting anaerobic (low oxygenating) conditions in contrast to the strong oxidising conditions of the red playa sediments. The sea was starting to arrive again.

Above the Tea Green Marls there is a dramatic change in the rock sequence. Dark shales appear, with marine fossils marking the base of the

Rhaetian stage, the final stage of the Triassic period. In southern Europe where all the Triassic sequence is marine, the Rhaetian is the last stage of the Triassic based on its fossil content. A shallow sea spread over the previous desert of north west Europe. Bone beds occur with vertebrate remains, often on a bored surface of Tea Green Marls. The marine transgression occurred almost simultaneously over Europe, from the Alps to the Baltic and across to Britain, and the sea spread rapidly over the low-lying, arid land surface. In Britain the Rhaetic sequence is now named the Penarth Group and includes marine fossils — oysters, scallops and sea urchins. The bone beds contain bones, teeth and scales of both marine and non-marine vertebrates.

The Rhaetic beds in Shropshire only occur in the north, above the Tea Green Marls around the Jurassic areas of Prees, and are entirely drift covered. In boreholes the thickness is estimated at 12-14 m. East of Whitchurch boreholes have proved 8 m of fossiliferous dark grey shales and mudstones overlain by 4 m of pale grey calcareous mudstones, sometimes called the 'White Lias'. North east of Wem boreholes prove 6 m of the *Avicula contorta* Beds overlain by 5 m of White Lias.

This rapid marine transgression, with similar beds found over the whole of Britain, marks the return of true marine conditions for the first time since the formation of the Carboniferous Limestone 330 million years ago. The Rhaetic Beds are about 205 million years-old and thus, for over 120 million years, Shropshire and most of Britain experienced continental conditions and often very arid, desert environments.

This marine transgression, caused by a branch of the Tethys Ocean spreading north west, was to continue into the next geological period, the Jurassic, and warm seas swarming with ammonites and other shell fish, as well as the giant marine reptiles, the ichthyosaurs, spread over Britain. Unfortunately hardly any Jurassic rocks are preserved in Shropshire, as we shall see.

Mineralisation in the Triassic rocks of Shropshire

Copper mineralisation is well known in north Shropshire and Cheshire within the highest Wilmslow Sandstone Formation (the old Upper Mottled Sandstone), and particularly the basement beds of the overlying Helsby Sandstone Formation (Keuper Sandstone). Disused copper mines are particularly well known at Clive (on Grinshill), Pim Hill, Yorton Bank, Wixhill, Hawkstone and Eardiston near West Felton. Barytes is also associated with the copper but this mineral is also found more widely disseminated within the Keuper Sandstone (Grinshill Sandstone) around Grinshill and elsewhere.

Around Clive and Grinshill the copper veins in the Grinshill Sandstone contain green malachite and blue azurite and are clearly related to north-south faulting. This is also true at Alderley Edge in Cheshire. In Shropshire this faulting is post-Lower Jurassic although the age of mineralisation is clearly younger than this, being probably of Tertiary age, and is about fifty million years old. Igneous activity of this age is present in Shropshire, and formed dolerite dykes around Grinshill to be described later, as well as chemically altering the Grinshill Sandstone to its present colour. This igneous activity was associated with the initial formation of the north Atlantic when North America moved away from Europe. Ascending magmatic fluids rich in copper, and with some barytes, were stopped from reaching the surface by a cap of impermeable Keuper Marl, and the minerals crystallised out in the underlying sandstones. At Grinshill the Tertiary dolerite dykes themselves contain clusters of barytes crystals.

Just west of Clive Church a number of shafts and galleries follow veins rich in copper carbonates, and one vertical vein was extracted to a width of 6 m. The Clive mine was abandoned by 1900 but seventeenth century tools found in old levels show that mining continued over a long period.

The Eardiston mine was worked between 1827 and 1865. Copper ores, mainly malachite, taken in 1841 yielded up to twenty-five per cent of their weight of copper, and 2,500 tons of copper ore was extracted in 1841-1845. The vein here was 1-2 m wide with granular quartz, mammellated malachite, and barytes. Barytes which occurs in the veins and also farther afield can be seen as star-shaped clusters in the sandstones on the top of Grinshill.

Chapter 10
Pieces of the Jigsaw missing — the Jurassic, Cretaceous and Tertiary periods

The main way in which a geologist understands the past is through a study of the rocks of a particular period. If no rocks of a geological period are preserved in the area being studied one can only deduce what might have happened by comparison with neighbouring areas.

This is the case in Shropshire for a large interval of time from about 190 million years ago to about only 100,000 years ago, since we have no rocks in Shropshire which belong to the Cretaceous or Tertiary periods (except some dykes around Grinshill), and only the lowest part of the Jurassic system is represented by rocks. So our records of consolidated rocks, what geologists call the solid rock formations, finishes in Shropshire with shales and limestones of the Lower Jurassic Lias around Prees formed about 180 million years ago. The next sediments we find are not really compact, solid, rocks but unconsolidated sediments which we call superficial or drift deposits, and the oldest of these in Shropshire are about 100,000 years-old, formed during the 'Ice Age' of the present Quarternary period which will be described later.

However, we can postulate what was happening over Shropshire during this long interval of time by looking at the Jurassic rocks of the Cotswolds; the Cretaceous rocks of Southern England, particularly the chalk of the Downs; and the Tertiary rocks of both south east England and north west Scotland.

The Jurassic period

The Jurassic period lasted from 205-135 million years ago. It was named after the Jura mountains on the France-Switzerland border, where the rocks are best displayed and were first studied. Over Britain we saw the expansion of the shallow seas of the late Triassic to form marine deposits over most of Europe and Britain, and then at the end of the period the sea over Britain retreated to the south coast area.

This shallow sea, a branch of the Tethys Ocean which separated Europe and Asia (Eurasia) from Africa, India and Australia at that time, teemed with both invertebrate and vertebrate marine life, particularly ammonites (Fig. 52), and on land the dinosaurs reached the height of evolutionary diversification.

Unfortunately, in Shropshire we only have a small area of Jurassic rocks, poorly exposed, around Prees in North Shropshire (Fig. 4). Modern boreholes have shown the drift-covered Lower Lias to have a larger outcrop than that shown on Geological Survey Sheet 138 (Wem). The outcrop extends a further 3-5 km west and south west than shown on that map, and I have amended Figs. 4 and 5 to show this. The town of Wem is probably underlain by Jurassic rocks. The rocks belong to the lowest division of the Jurassic system known as the Lias. This latter term is an old Dorset word to describe the alternations of shales, clays and thin limestones, occurring as layers (or Lias?) on the coastal cliffs of Dorset, and Yorkshire, which are famous for their fossil ammonites and giant marine reptiles, the ichthyosaurs. Because the British Jurassic rock sequence is entirely marine, we have no fossils of land dinosaurs of this age. You have to go to North America or eastern Europe to find these.

The Lias is divided into three divisions, Lower, Middle and Upper, and the lower two occur around Prees, although the base of the sequence and the Lower Lias are covered by glacial drift. Boreholes have proved at least 160 m of Lower Lias, comprising grey-blue shales, clays and thin limestones resting conformably on the Rhaetic Beds. The true thickness of the Lower Lias is probably nearer to 300 m as the boreholes started in the middle of the sequence. At Prees a small but prominent hill is formed by the Middle Lias with exposures in 30 m of micaceous sandy marls, calcareous mudstone (marlstones) and sandstones, and grey shales. Fossils (Fig. 52) are common, many of which are preserved in Shrewsbury Museum and include: ammonites; belemnites (ancestors of the squid); bivalve molluscs, including oysters and scallops; and brachiopods.

During 1988 the construction of the Prees by-pass exposed a good section of the Middle Lias. Grey shales passed up into about 6 m of very dark grey, well laminated shales, often with small marcasite nodules, and with belemnites and large bivalve molluscs. Some calcareous layers were packed with belemnites and some small patches of jet occurred. A small bench marked the level of an overlying harder layer of ferruginous marlstone (a marlstone is a hard calcareous mudstone) up to 1 m thick, and above this level about 5 m of brown and greyish-brown calcareous shales and thin, grey, sandy limestones occurred. This level contained belemnites, bivalve mulluscs and abundant ammonites. The harder marlstone in the middle of the exposed sequence probably equates with the well-known Marlstone Rock Bed of the north Cotswolds. In the early stages of the by-pass construction, it could be seen how the marlstone formed a small but prominent bench on the slopes of Prees Hill.

Here the rock sequence finishes and we can only suggest that by looking at the remaining Jurassic rocks of the rest of the country, and the suggested

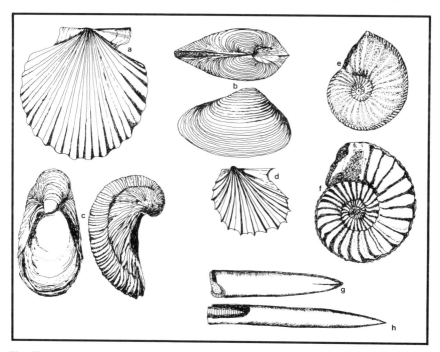

Fig. 52. Jurassic fossils from the Middle Lias of Prees. a. *Pseudopecten*, x0.5; b. *Pleuromya*, x0.7; c. *Gryphaea*, x0.5; d. *Oxytoma*, x0.7; e. zonal ammonite, *Amaltheus margaritatus*, x0.3; f. zonal ammonite, *Pleuroceras spinatum*, x0.5; g., h. belemnites, x0.5 and x0.3.

ancient geography for the time (palaeogeography), the rest of the Lias was deposited over Shropshire, followed by the famous oolitic limestones of the Cotswolds and various clays, such as the Oxford Clay, and further limestones, before the sea retreated to the south towards the end of the period. Late Jurassic rocks such as the Kimmeridge Clay and the Portland Stone were never laid down over Shropshire. However, the earlier formations certainly were laid down and have now been removed by erosion so that just a small patch of Lias remains in the north of the county.

The Cretaceous Period

The Cretaceous period, named after the Latin for chalk, lasted from 135-66 million years ago and is famous over southern and eastern England for its deposits of the Chalk, a remarkably uniform and pure, soft, white limestone sequence. The Chalk was almost certainly laid down on the floor of a shallow clear sea, at around 40°N. Under a modern electron microscope it is seen to be made up almost entirely of millions of minute coccoliths.

Coccoliths are minute calcareous scales, three to fifteen microns in diameter, forming a protective armour to a unicellular spherical coccolithophore, which is up to twenty microns in diameter. These are primitive planktonic plants belonging to the group called protista. They appeared in the Triassic, are abundant today, but were particularly abundant in the Cretaceous. The coccolithophores rained down in their millions as they died, and fell to the sea floor to build up Chalk from the individual coccolith plates as the sphere broke up.

The Chalk is the youngest of the Cretaceous rock formations in Britain. Earlier formations include the Wealden Sands and Clays found in Kent and Surrey in which the new flesh eating dinosaur 'Claws', *Baryonyx walkeri* was discovered in 1983. These earlier Cretaceous formations were only deposited in south east England as deltaic deposits fringing a northern continent with tropical forests. However, later subsidence allowed a shallow sea to spread over almost the whole of Britain, and in this the Chalk was deposited. Although today it is best known from southern England, it is also found in Northern Ireland and a small outcrop occurs, preserved in a volcanic vent on the Isle of Arran. Drillings have proved it under the Irish Sea, and it would seem probable that it also covered the whole of England, including Shropshire, and has since been removed by erosion.

The Tertiary period

The Tertiary period was one of great importance in north west Europe in terms of earth movements (tectonics). It lasted from 66 to 2 million years ago and during this time two important tectonic events happened.

Firstly, about 40 million years ago, the northward movement of Africa towards Europe caused the Alpine orogeny and the formation of the Alps themselves from the sediments within the western part of the Tethys Ocean. These earth movements died away northwards, but in southern

Britain were powerful enough to cause quite tight folding of Cretaceous and older strata in Dorset and the Isle of Wight. The well-known folds of the London Basin and the Weald anticline also date from this period.

Secondly, up to the end of the Cretaceous period, Britain and Eurasia had been firmly attached to Greenland and North America, forming the northern part of Alfred Wegener's Gondwanaland, which he called Laurasia. During the early Tertiary, North America and Greenland started to move away from Eurasia and the north Atlantic Ocean began to form, with the mid-Atlantic ridge at its centre with its volcanic activity causing sea floor spreading and the formation of new basaltic oceanic crust.

The initial split, about fifty million years ago, took the form of rifting between Greenland and north west Scotland and violent volcanic activity broke out in these areas, causing great areas of basalt lavas to cover Northern Ireland, the Antrim plateau, including the classic basaltic lavas of the Giant's Causeway with their regular hexagonal columnar joints. These basalt lavas are also found extensively in Skye, Mull and on the Isle of Staffa, where Fingal's Cave shows columnar jointing similar to that across the water at the Giant's Causeway.

Volcanoes, now eroded away to their roots, occurred on Mull and Arran, and vertical thin sheets of igneous rocks, called dykes, were injected into the surrounding rocks. These are often so numerous as to occur in what are called dyke swarms. Individual dykes travelled large distances and those found in north Shropshire around Grinshill are of Tertiary age, as will be explained later. All volcanic activity has now transferred to the main mid-Atlantic ridge, which is causing the Atlantic to spread at about 2 cm per year around Iceland.

The splitting of any continental mass to form a new ocean in between is preceded by a doming up of the area where the crack will first appear, and then the formation of a major rift valley system in this domed-up area, along which volcanic activity will break out as the convection cells from the mantle bring volcanic material to the surface. The classic example of this today is in East Africa where the major rift valley in a domed-up highland area has extensive volcanic activity and is probably the site of an attempted split in the African continent. Whether it will continue to happen is debatable, since Africa itself is being squeezed by the spreading Atlantic and Indian Oceans on either side.

The initial doming up of a continental area prior to splitting occurred in Britain in the Tertiary, since all of Britain experienced a gentle tilt from north west to south east which produced, or to a large extent re-exhumed, the ancient highlands of the north-west. The Cretaceous sea which had laid down the Chalk was forced to retreat to the south to cover only parts of south east England, and so we can say that during the Tertiary the British Isles started to take on the shape of its present landsurface, although the coastline was very different, as we were still joined to Europe. The highland areas had yet to be worn down by the glaciers and ice sheets of the Ice Age, and the lowland areas had yet to receive their covering of superficial (drift) deposits caused by the passage and retreat of the ice sheets.

By the end of the Tertiary, Britain and north west Europe, as well as splitting away from North America, had drifted northwards to its present latitude. During the Tertiary the climate over north west Europe was much warmer than today in temperate latitudes, and this led to a good deal of quite deep weathering on the recently exhumed landsurfaces.

However, the close of the Tertiary period, about two million years ago, and the start of the present Quaternary period, in which we now live, was heralded by a marked global cooling of the climate which led to the great Ice Age of the last one million years or so in the areas adjacent to the polar regions.

The Tertiary period in Shropshire

There are no sedimentary rocks of Tertiary age in Shropshire and so we can only speculate that after the retreat of the Cretaceous Chalk Sea the land rose above sea level as part of the tilting process described above. Erosion took place to remove the Chalk and most Jurassic rocks and to form eventually the distribution of highland and lowland areas that we see today. Landforms had much more rounded outlines, and several different drainage directions, for instance, the Severn flowed north to the Dee estuary rather than eastwards through the Ironbridge Gorge which had yet to form.

The Alpine orogeny, which caused folding in southern Britain, affected Shropshire to a lesser degree. The Triassic-Jurassic basin of north Shropshire is a post-Jurassic fold structure, and faults which cut Jurassic strata such as the Prees or axial fault (Fig. 5) are almost certainly Tertiary faults. Many earlier faults, including the Church Stretton Fault and Longmynd Scarp Fault, were reactivated and had post-Triassic movements along them, and these are probably of Tertiary age.

The Tertiary igneous activity of north west Scotland referred to earlier resulted in a few dykes reaching as far south as the Midlands, and Tertiary dolerite dykes occur around Grinshill, cutting Triassic sandstones. They were discovered by Murchison, who refers to them in 1835 around Grinshill and Acton Reynald House. The dolerite is much altered, and on Grinshill the dyke is up to 1 m thick, and can be seen in the present working quarry. Four branches occur around Acton Reynald, varying from 1-3 m in thickness. Radiometric dates give an age of 51-52 million years.

The dykes appear to have later barytes mineralisation, and may be themselves affected by north-south faulting of late Tertiary age.

This episode of igneous activity may well have caused the widespread copper mineralisation around Clive and elsewhere in north Shropshire, and also chemically altered the Grinshill Sandstone around Grinshll and Hawkstone, so that it now has a dominantly pale colour and is harder than the red sandstones of the same age around Nesscliffe and Ruyton.

This chapter has shown that quite a lot can be deduced about conditions over Shropshire during the Jurassic, Cretaceous and Tertiary periods, although hardly any sedimentary rocks of the periods remain.

Chapter 11
Towards the Present Day — the Quaternary Period, including the Ice Age

Today we are living in the Quaternary period which started about 2.5 million years ago with a major worldwide climatic change to cooler conditions which heralded the onset of the Ice Age. The Quaternary period is divided into two epochs, the Pleistocene (often incorrectly equated with the term Quaternary, see Table 1) and the Holocene or Recent, which started only 10,000 years ago after major retreats of the ice sheets. So we are now living in the Holocene epoch of the Quaternary period, and in a few thousand years' time another climatic change, or some other event, may make geologists decide to start a new epoch. The onset of the Ice Age was almost certainly caused by plate tectonic movements resulting in the Antarctic continent becoming firmly established over the south pole and the northern continents moving closer to the north pole. This allowed the formation of continental ice sheets, and once these started to form, maybe as early as five million years ago in the arctic areas, the ice in turn affected the climate, enhancing the cooling effect. The first extensive glaciation of the North Atlantic region occurred 2.4 million years ago, but we cannot be sure whether this glaciation affected the British Isles or not. However, during the subsequent period there have probably been at least twenty cold, glacial periods, all of which have caused advances of ice sheets from the north, followed by warmer interglacials. Variations in the Earth's orbit around the sun, the Milankovitch theory, explains how we have had fairly regular cooling and warming of the climate at intervals of about 120,000 years (Brunsden *et al.*, 1988). Antarctic glaciations may have started as far back as twenty-two million years ago at the start of the Miocene epoch of the Tertiary period.

The evidence for these twenty cold periods in the last 2.4 million years comes from oxygen isotope ratios in the carbonate content of minute calcareous marine organisms found in deep sea cores (Brunsden *et al.*, 1988, Fig. 1.4). Whether all these cold periods caused glaciations over Britain is difficult to say, as there is no evidence on land in the form of glacial sediments or features.

In Britain we only have detailed evidence on land of the last three glacial periods, starting with the Anglian glaciation which started about 270,000 years BP (before present). This was the most extensive glaciation and the ice spread as far south as a line from London to Bristol, and then west along the North Devon coast to the Scilly Isles, whose northern coasts show smooth glaciated surfaces. The Wolstonian glaciation is dated at

Table 9. Stages of the Pleistocene Epoch.

	Stage in Britain		Climate	Glacial Stages of N.W. Europe	Alps
QUATERNARY	Holocene	Flandrian (post glacial)	temperate		
	Pleistocene	Devensian — 10,000 years BP / 80,000 years BP	cold, glacial	Weichselian	Würm
		Ipswichian — 120,000 years BP	interglacial		
		Wolstonian — 200,000 years BP	cold glacial	? Saalian	Riss
		Hoxnian	interglacial		
		Anglian — 270,000 years BP	cold, glacial	? Saalian ? Elsterian	Mindel
		Cromerian	temperate		
		Beestonian	periglacial and glacial		Günz
		Pastonian — 0.6 to 1.6 million BP hiatus	temperate		
		Baventian	cold		
		Antian	temperate		
		Thurnian	cold		
		Ludhamian — 2 to 2.5 million BP hiatus	temperate		
		Waltonian — 2.5 million years BP	temperate		
	Pliocene		warm/ temperate		

between 200,000 and 150,000 years BP, and may have been as widespread as the Anglian. There then followed a warm interglacial from 120,000 to 80,000 BP, after which the Devensian glaciation started and reached its greatest extent at only 18,000 years ago. Since then, the climate warmed rapidly and ice had disappeared from southern Britain by 13,000 years BP. There was a minor re-advance of the ice in Scotland, the Loch Lomond advance, between 11,500 and 10,800 BP. Reconstructions of the thickness of the Devensian ice suggests a maximum of nearly 6,000 feet over southern Scotland, and this was probably greater during the Anglian and the Wolstonian. The Devensian glaciation reached as far south as the two earlier glaciations. However being the last (for the time being at least!) its deposits are well exposed.

Of course, the Quaternary includes events which happened all over the world, as do all geological periods, and so we cannot call the period 'the Ice Age', since cold glacial conditions only affected parts of the world near to the polar regions, and highland areas elsewhere. As we have seen earlier, referring to the Ice Age in the singular is wrong but we nearly always do this. During the Pleistocene, between the major advances of the ice from the north, we had interglacial phases when the climate was often warmer than it is today (Fig. 53), and all ice disappeared over Britain.

In Shropshire we would say the Ice Age has finished, and indeed we often refer to the Holocene as the Post Glacial, but if you are living in Greenland today you would hardly agree. Similarly, the equatorial regions never experienced an Ice Age and here people can only refer to Quaternary sediments and their contained fossils reflecting a cooling of the climate. During the coldest phases of the Devensian glaciation, when much of Britain, including Shropshire, was covered by thick ice sheets as recently as only 18,000 years ago, you could still have had quite a warm holiday in the Caribbean. Average temperatures in the Atlantic areas were probably 8°C lower than today.

The start of the Quaternary around 2.5 million years ago saw a cooling of the Earth's climate so that polar ice caps appeared and started to spread south and north. At the height of the Ice Age the polar ice sheet reached as far south as the Thames valley and the Scilly Isles in Britain, into central Europe and well south in to the USA. The southern ice spread north from Antarctica into the surrounding oceanic area but did not reach Africa, Australia or New Zealand, but did reach the southern parts of South America. As well as polar ice sheets, highland areas throughout the world, including those in equatorial regions, produced ice sheets and small ice caps with valley glaciers spreading from them. These were far more extensive than they are at present, and in places such as East Africa (Mount Kenya and Ruwenzori, etc.) and New Guinea you can find evidence of the much wider extent of valley galciers and ice sheets, of which only small remnants exist today.

All this, remember, started from 'zero' with no ice at the poles or, indeed, anywhere during the Tertiary period. Ice at the poles is a rare geological event, not the norm.

The continental ice sheets of Greenland, Antarctica and elsewhere were produced by precipitation falling as snow and being compacted to ice. These continental ice sheets thus locked up water which had evaporated from the sea to cause the snowfall and thus sea level fell throughout the world during the Ice Age. These areas must not be confused with the frozen Arctic Ocean around the North Pole which is simply frozen ocean, not an ice sheet.

Now that the continental ice sheets have been continually melting for the last 15,000 years, sea level has been rising to a level which was normal during the last 500 million years or so. Indeed, if the world's climate continues to warm up and all the continental ice melts, sea level will rise worldwide by about fifty metres, with catastrophic effects on coastal areas and cities. In the last 15,000 years this rise in sea level has been partly offset in the glaciated areas by the rise or 'bounce back' effects of continental crust now released from the enormous weight of overlying ice. This isostatic readjustment, as it is called, has caused parts of central Scandinavia to rise by up to 300 m, having been depressed by that amount during the height of the Ice Age under an estimated maximum thickness of 2,650 m of ice. Northern Britain has also experienced this isostatic rise in the last 15,000 years, and is still rising by about 3 mm per year, but in the south of the country around London, which never had any ice cover, there cannot be any readjustment and so the full effect of any rise in sea level will be felt.

This rise in sea level will almost certainly happen, unless the climate cools again and the ice returns, and modern opinion, as has been said earlier, suggests that ice at the poles is not the norm, and that only four Ice Ages have occurred in the last 700 million years: one about 650 million years ago in the late Precambrian, one in the late Ordovician 450 million years ago, one at the end of the Carboniferous 300 million years ago, and that of the present Quaternary.

The most famous of the older Ice Ages is that of the late Carboniferous-early Permian, often called the Gondwana glaciation because it affected all the southern continents which were at that time clustered around the south pole — Antarctica, South Africa, South America, Australia, New Zealand and India — and which Alfred Wegener called Gondwanaland. It was the presence of undoubted glacial effects at that time in all these continents which was one of Wegener's major lines of evidence for continental drift. For, he argued, how could South Africa and India show evidence of simultaneous glaciation when they are now so far away from the poles, unless in the past they were in fact very close to the south pole and have since drifted away? We now know this to be the case — India has drifted 14,000 km northwards in the last 150 million years.

It is difficult to estimate how long Ice Ages have lasted for, our dating techniques for rocks millions of years old do not allow us enough accuracy to date to more than say ±5 million years. During the 'present' Ice Age, radio carbon dating allows much better accuracy of say ±300 years in 12,000 years. The Gondwana glaciation certainly lasted longer than our present Ice Age, but has this, in fact, finished? There is perhaps no reason why any Ice Age should be of a particular length. The ice sheets are retreating now, but whether the Quaternary Ice Age has finished, or is about to return in a few thousand years is a matter of great debate. Are we

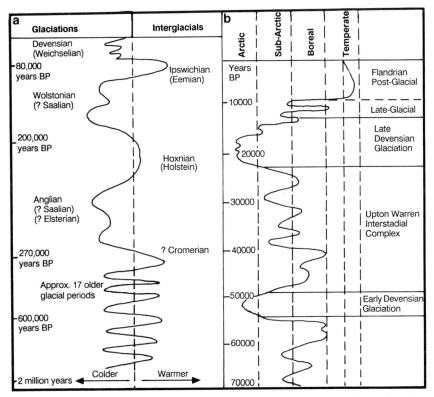

Fig. 53. Climatic fluctuations in the Pleistocene. Main variations on the left showing broadly the cold and warm periods. The three most recent and well documented British glacial periods are shown, but it is now accepted that up to twenty advances and retreats of the ice sheets may have occurred during the Pleistocene epoch of the Quaternary period. Detailed variations of the last 70,000 years are shown on the right. (After Coope and Sands, 1966, Proc. Royal Soc., B, Vol. 165, p. 389.)

in an interglacial now or just a short warm period? The present distribution of the continents and the Milankovitch effect explained earlier suggests that cold-warm-cold cycles will continue well into the future, and so the Ice Age has not finished, unless Man interferes with the climate enough to cause 'unnatural' climatic changes.

Following the onset of the Ice Age in the early Pleistocene, the ice sheets advanced south over Britain and Europe up to twenty times, although only the last three glacial periods can be recognised with certainty in Britain. The glacial periods were separated by warmer interglacials and all these cold and warmer stages have been given names. In the early 1900s the four major glaciations of the Alps were given the famous (or infamous) classic names of Günz, Mindel, Riss and Würm in order of age, Günz being the oldest. Geologists elsewhere in Europe have found these terms difficult to apply; this is certainly the case in Britain where we have evidence for only three glaciations, termed the Anglian, Wolstonian and Devensian (the most recent). The various stages and their probable correlations are shown in Table 9.

Fig. 53a shows graphically the temperature variations during the Quaternary, and the three 'cold' peaks show quite clearly the most recent glacial episodes. One can see how warm the last interglacial (Ipswichian) was compared with today. Fig. 53b is a more detailed graph for the last 70,000 years and shows variations within the Devensian.

How are these climatic and temperature variations deduced? They are mainly based on the plant and animal remains found in the associated glacial and interglacial sediments, particularly pollen and beetles. The trees, plants and insects which lived during the last two million years are similar, and in many cases identical, to those living today. Thus abundant dwarf birch indicates a cold climate, and abundant oak and ash indicates a temperate climate. Beetles inhabit all environments from warm to glacial, and thus the conditions in the Devensian were cold enough to allow beetles which live in the arctic today to inhabit Britain 30,000 years ago. Pollen analysis and the study of beetles has allowed Quaternary geologists to deduce climatic variations very accurately, as shown in Fig. 53b.

Glacial deposits and physical features caused by glaciation

Deposits of glacial origin, like all others, take characteristic forms. The most well known is called Boulder Clay, or more correctly, Till. Boulder clay is a rock flour produced by the grinding action of ice sheets and glaciers as they pass over the solid bed rock. The ice also picks up boulders and rock fragments which it uses at its base as grinding tools, and these boulders themselves can eventually be reduced to flour. When the ice retreats it leaves behind a mixture of clay and boulders, hence the term boulder clay. The boulders can vary in size but are usually well rounded, smoothed and occasionally marked with scratches. Many boulders have travelled considerable distances from their original source where they were first picked up by the ice, and are called glacial erratics. In East Anglia one can find erratics from Norway, and in Shropshire many different granite erratics have come from Scotland and the Lake District.

Till is the term used more often nowadays than boulder clay. Frequently it is sandy, or a mixture of sand and clay. Clay tills (true boulder clays) are sticky, blue clays (orange when weathered) which can be a gardener's nightmare in parts of Shropshire. As well as glacial tills, melting ice sheets and glaciers produce large amounts of fluvio-glacial sands and gravels which are washed out of the base and the edge of the ice by melt water streams. Kames are masses of gravel dumped at the edge of an ice sheet and Eskers are sinuous ridges of sand and gravel formed in tunnels within the ice by melt waters flowing between the ice and the ground surface. In Shropshire a well-known esker, the Dorrington esker, forms a conspicuous ridge curling from near Lyth Hill south towards Dorrington.

Glacial moraines are formed on the top of a glacier as material falls on to its surface from the valley sides, and also at its snout. When the ice melts they are left behind on the ground surface over which the glacier or ice sheet has been moving. Moraines are usually irregular dumps of material and will be reworked by melt streams. Examples of morainic gravels are found in Shropshire.

Drumlins are well-known land forms left behind by moving ice sheets. Sands and gravels deposited by earlier glaciations are further moulded into smooth elongated hillocks with the blunter end towards the ice. They have the shape of half an egg, typical dimensions being 400 m by 100 m, and often occur in large numbers in drumlin fields and produce a type of hummocky land surface called 'a basket of eggs', well known in northern England. Drumlins are rare in Shropshire, but the hummocky floor of the Church Stretton valley towards All Stretton has a look not too dissimilar from a drumlin field.

Quite a famous glacial landform common in Shropshire is a Kettle Hole, and it was in one of these (Fig. 54) that the famous Condover mammoths met their deaths. When an ice sheet, such as that which covered Shropshire north of the south Shropshire hills about 18,000 years ago, is melting and 'retreating', masses of ice can become isolated from the main ice sheet and become stagnant blocks of ice surrounded and covered by outwash sands and gravels, and clays. When this mass of ice finally melts, the sediments collapse into the large hole left behind, giving rise to a depression called a kettle hole. These have no external drainage and can fill with water to form a small lake with steep sides, in which clay and peat can accumulate. These small lakes are called Meres in Shropshire, and will be described later. If the lake completely fills with clay and peat it becomes a bog or moss, such as Wem Moss in north Shropshire.

Much larger lakes can form at ice fronts between the ice and any high ground or enclosing ridge. This glacial lake can receive seasonal deposits of clays with beautiful banding called varved clays. Glacial lakes will increase in size and may eventually overflow surrounding high ground, forming glacial overflow channels. The most famous of these in Shropshire is the Ironbridge gorge which will be described later. Large masses of ice can also block off normal directions of river flow causing glacial lakes to form, which overflow and create new river channels which the rivers follow, even when the ice has retreated. Major river diversions of this type are common in Shropshire, but again, that of the River Severn is the most famous, and will be described later.

All the features described so far are characteristic of a lowland area such as much of north Shropshire. Mountainous areas show characteristic glacial features such as: 'U' shaped valleys; striated rock surfaces; roche moutonées; arêtes; and corries or cwms, with their small lakes, which can be seen in North Wales, the Lake District and Scotland. They are not, however, seen in Shropshire, not even in the south Shropshire hills, which were not high enough to produce their own ice caps.

Periglacial Landforms and Deposits

Areas adjacent to or above ice sheets during the Ice Age were affected by what is called a periglacial climate, a climate which can be studied today in polar regions. Here, extreme cold at night can be followed by relatively high temperatures in the day. This causes 'freeze-thaw' action when water which has been trapped in joints of crags expands at night, causing the rocks to become shattered as the process is repeated day after day. Large scree fields form around the frost shattered crags which are called tors. The Stiperstones is a classic area for these features.

In areas where bare rock does not occur at the surface, freeze-thaw action causes the whole hill surface to become broken up to a depth of a few metres, into a coarse angular deposit called Head. Large areas of ground are affected by very slow gravity movements of these fractured surface deposits, a process called solifluction. This type of process is seen to good effect on the top of the Longmynd, along with large areas of head.

The Ice Age in Shropshire

Although the Anglian glaciation must have affected Shropshire, the evidence for it has been destroyed by late glaciations and erosion. Ice of

Wolstonian age probably spread from Wales eastwards to leave sediments in the Onny Valley south of the Longmynd and as far north as Marshbrook.

The last, or Devensian, glaciation had profound effects on Shropshire and the evidence for it is widespread. This is the best documented glacial period since its deposits are abundant, and in particular have not been reworked by later glaciations.

After the warm Ipswichian interglacial, the Devensian cold phase began again 80,000 years ago, and ice sheets spread south from the Irish Sea area over the Cheshire Plain, into the north Shropshire Plain and reached as far south as the Church Stretton Valley at Little Stretton. The Devensian glacial maximum is now calculated to be as recent as only 18,000 years ago, with sea level depressed by up to 100 m lower than today, and with up to 1,800 m of ice over Scotland. Evidence from boulder clay found at heights of 300 m O.D. on the northern Longmynd suggests that the ice must have been that thick over north Shropshire, and boulder clay occurs at similar heights on the sides of Church Stretton Valley. Other authors (Brunsden *et al.*, 1988, Fig. 1.6) suggest thicknesses of Devensian ice in Shropshire in excess of 1,000 m. Ice from the Welsh Mountains spread east to meet the Irish Sea ice in the Oswestry and Shrewsbury area, and also spread into South Shropshire via the Clun and Teme Valleys and from the Church Stoke area.

Periglacial Formations
The high ground of the South Shropshire hills, above about 300 m, protruded as nunataks above the ice, but was affected by a periglacial climate. The frost shattered tors of the Stiperstones, now preserved as a National Nature Reserve, show marvellous examples of freeze-thaw shattering of the rocks (Plate 15). The widespread screes show stone stripes and polygons. The north slope of Titterstone Clee has large screes of dolerite formed in the same way, while the Longmynd plateau has extensive head deposits.

Glacial deposits and landforms
The Church Stretton valley, prior to the arrival of ice, was deeper than it is today by about 30-60 m. Its bottom is now full of sand, gravel and boulder clay left after the retreat of the Devensian ice. The ice maximum was probably about 18,000 years ago, after which the ice began to melt and retreat northwards, so that by 13,000 years ago there was no ice left in Shropshire. The glacial sediments and features of this retreat phase are the most important point to discuss.

I have been continually referring to ice retreating, but this never means that ice sheets or glaciers ever more backwards, i.e. uphill, or northwards. By retreat I mean that the ice front was melting quicker than supplies could be replenished from the north or high ground. Thus the ice front appears to be retreating although the motion within the ice is always away from the source. Perhaps waning is a better term than retreating.

Boulder Clay (Till)
Boulder Clays up to 100 m thick cover large areas of north Shropshire and occur as far south as the north Longmynd and Church Stretton Valley. Ice does not seem to have passed over Wenlock Edge as boulder clay is absent in Corvedale and the Clee Hills. Erratics in the boulder clay come from all over North Britain and characteristic granites from Scotland and the Lake District can be found, as well as rock from more local sources.

Fluvio-glacial sands and gravels, and associated land forms
These cover large areas of the north Shropshire Plain and areas east of Telford, as well as south of Shrewsbury in the Dorrington-Condover area, where they are extensively worked for sand and gravel. Similar deposits occur in the south of the county in the Clun, Teme and Onny valleys associated with the Welsh ice.

These are all melt water sands and gravels, formed as kames and eskers and terminal moraines. A very well-known esker ridge has already been referred to north of Dorrington towards Lyth Hill and, east of the Cound brook, rounded gravel and sand ridges occur around Condover. In the Condover area a number of kettle holes formed, some of which, e.g. Bomere Pool, are still full of water, and are very steep sided. Others were completely filled with clay and, later, peat so that they appear only as small, badly drained depressions in the hummocky terrane. One of these, only 800 m west of Bomere Pool, was discovered in 1986 during quarrying for gravel, and during removal of the clay and peat to get at the underlying gravel, the Condover mammoths were discovered.

The most well-known kettle holes in Shropshire are in the north of the county, many of which contain water — the famous meres of north Shropshire. The area around Ellesmere is an ice moulded landscape of sand and gravel hillocks with clay lined hollows. Some of the kettle holes filled with clay and peat and became the mosses of the area, such as Wem Moss. Whixall Moss is a very large area of peat and can be considered as having been formed when a large mass of ice became isolated from the main ice sheet. Classic kettle holes are usually smaller, such as Colemere and Blakemere, but with quite steep sides in places.

The Condover Mammoths
One evening in September 1986 a remarkable discovery was made by Eve Roberts of Bayston Hill, near Shrewsbury, who was walking her dogs round the edge of a gravel pit at Norton Farm just west of Bomere Pool, Condover (Plate 30). She noticed large bones sticking out of a pile of clay

and peat which had been dumped on ground adjacent to the gravel pit that very day. The story from now on is perhaps well known to local people. The County Museum Service was contacted by Eve Roberts and they collected a large number of bones the following day. Dr. Russell Coope from Birmingham University, later joined by Dr. Adrian Lister from Cambridge University, examined the bones and concluded they were from an adult woolly mammoth.

The owners of the gravel pit, ARC, gave financial and other help, and over the next 18 months the large pile of dumped material was re-excavated and sorted. This work yielded numerous remarkably well preserved bones of an almost complete adult woolly mammoth, except for its skull and tail. It appears to have been in good health when it died. Even more exciting was the discovery of parts of three young mammoths, including each of their lower jaws, two of which were four years-old and one six years-old. Examination of the jaw and teeth of the adult showed it to be using the fifth of its six teeth, and thus by comparison with modern elephants was thirty to thirty-two years-old when it died. Radio carbon dating of the bones gave an age of 12,700 years for this unique find.

When the adult died, the epiphyses of the bones showed it to be still growing. Female elephants stop growing in their mid-twenties but males continue to grow, and as this adult was thirty to thirty-two years-old, and still growing, it was considered to be a male. Russian scientists who examined the pelvis also concluded that it was a male. Two of the juvenile pelvises showed marked differences which suggests one is male and one female. About five years before the adult died, its shoulder blade had been fractured by a strong blow, and then healed perfectly. This damage was probably inflicted during fighting with other males, and did not result from an attacked by stone-age Man (late Palaeolithic) of whom we have no evidence at this time in Shropshire.

The mammoths appear to have been in good physical condition, with no evidence of any disease. The adult stood eleven feet tall (340 cm) at shoulder height, making it one of the tallest mammoths found in Eurasia. The mammoths are the youngest by 5,000 years to be found in Britain, and the assemblage of an adult and juveniles is unique in Europe. The adult skeleton is the best preserved mammoth of any age yet found in Britain. The last cold stage of the Devensian between 20,000 and 15,000 years ago did not therefore bring about the disappearance of the mammoth over Britain. They survived and returned in late glacial times. Other events in the late glacial caused the extinction of the mammoth in Europe.

The bones had been removed by diggers and dumped in a pile of clay and peat on ground adjacent to the gravel pit, and so were not found *in situ*. The gravel company had in fact come across a kettle hole containing peat and clay which was of no use to them, which was why they dumped it

The Quaternary Period, including the Ice Age 173

elsewhere. Bomere Pool, nearby, is another kettle hole still water-filled. So what was the original situation of the bones before they were removed from the kettle hole, and how did the mammoths come to be removed from the kettle hole, and how did the mammoths come to be in the kettle hole deposits? How were they trapped and fossilised?

The gravel pit was part of a larger area of hummocky ground of glacial sands, gravels and tills with enclosed hollows, being excavated for gravel and sand, and contained a kettle hole (Fig. 54) covering an area of two hectares (nearly five acres), and seven metres deep. At the time of the formation of the kettle hole, about 13,500 years ago, the original 300 m or more of ice over this part of Shropshire had probably thinned to around 20 m, as shown in Fig. 54. The floor of the kettle hole was a sandy till (boulder clay) with Welsh erratics, and the hole was initially filled with 4 m of pink laminated clay overlain by 1 m of blue-grey clay. At the margins of the kettle hole the sands and gravels dipped sharply at forty degrees

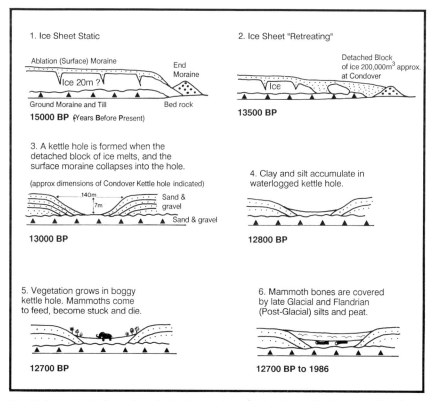

Fig. 54. Stages in the formation of a kettle hole, in particular that at Condover in which the famous mammoths were discovered in 1986. Dates are shown as follows: 15,000 BP equals 15,000 years before the present day.

forming steep sides to the hole. It would appear that at this stage the mammoths wandered into the steep sided pit, became stuck in the waterclogged clay bottom and could not get out. This was 12,700 years ago. However, Dr. Coope (personal communication) received dating information in 1989 which suggested that one of the juveniles may have died a few hundred years after the adult, which allows for various intriguing possibilities. After this, in post glacial (Flandrian) times, the hole was filled with 2 m of green-grey clay and peat, the peat containing well preserved red deer bones and one bear toe. Some of these deposits have been dated at 4,000 years-old. The mammoth bones were coated in blue grey clay which clearly showed they came from the 1 m layer, 4 m above the base of the kettle hole. It is a pity that no bones were found in *in situ*.

The dark clay surrounding the bones contained a rich assortment of fossil plants and insects. The juvenile skull and two juvenile lower jaws contained numerous blowfly pupae and dung beetles. This suggests that the animals died with the bodies below water, stuck in the clay, but with heads above water resting on mud and dung near to the side of the partly water-filled kettle hole.

The radiocarbon dates, stratigraphy and biological evidence suggest the mammoths lived 12,700 years ago during an early stage of a rapid climatic warming (amelioration) which followed the maximum Devensian glacial episode, and preceded a more recent, colder period at 11,000 years, which was then followed by the main post-glacial warming 10,000 years onwards. The presence of blowfly and dung beetles suggests a relatively mild climate.

The cold spell prior to the mammoths appearing in Shropshire was very intense, with average January temperatures of −20°C as evidenced by studies of beetles. Within perhaps a hundred years, the January temperature had risen to 0°C, and summer temperatures would have been around 10°C. Good conditions for mammoths. They were possibly migrating northwards when they were caught. The tundra vegetation over the area would have been sparse except in the lush damp hollows as found around Condover. The mammoths went down the steep sided kettle hole in search of good food and became stuck in the waterlogged, clayey bottom. There they died, after desperately trying to get out, as evidenced by the mixing of plants and insects in the clay deposits, caused by the thrashing around of these doomed creatures.

The Condover mammoths provide one of the great geological discoveries of this century. They again show us how fragmented is our knowledge of the record of animals left in the past. If ARC had not been searching for sand and gravel, we would not have known of the mammoths' existence in Britain during so recent a time. If Eve Roberts had not been walking her dogs, we might have no story to tell. Certainly, we would have no story to tell if the mammoths had not stepped down to feed at the bottom of a steep-sided kettle hole 12,700 years ago!

Glacial Lakes, including Lake Lapworth and the formation of the Ironbridge Gorge

Before the Ice Age in late Tertiary times, the early (proto) River Severn flowed down to Welshpool and then northwards to the present Dee estuary. With the coming of the Ice Age and the river drainage ceased to exist, but the valleys of the headwaters still remained, buried under the ice.

With the retreat of the Irish Sea and Welsh ice towards the end of the Devensian, 15,000 years ago, the upper Severn started to flow again, but at one stage the Irish Sea had only retreated from the southern part of the North Shropshire Plain and so the exit of the Severn to the Dee estuary was blocked. Glacial lakes began to form between the ice front and the high ground to the south, particularly against Wenlock Edge and the high ground between Telford and Newport. Eventually these lakes joined to form one large lake, which Professor Wills at Birmingham University named Lake Lapworth. The shoreline at its maximum extent lay 90 m above present sea level, and covered a large area of the southern part of the North Shropshire Plain from Newport, south west to Shrewsbury, west towards the Breiddens and north towards Wem with the sandstone hills sticking out above its surface, with the ice found to the north as its northern margin.

However, prior to the formation of Lake Lapworth we must consider the formation of smaller but important earlier lakes, which Professor Wills showed to have formed in the Telford-Newport area. In pre-glacial times two high level cols existed, one at Lightmoor near Madeley, and one at Ironbridge. Initially, an ice-dammed lake in Coalbrookdale overflowed at Lightmoor to cut the channel of the Coalport Brook to a great depth below the level (and to the east) of the Ironbridge col. The Lightmoor col was temporarily re-obstructed by ice so that the level of Lake Coalbrookdale was raised until the water lapped over the Ironbridge col. This water descended 50 m in just one mile from the Ironbridge col to the Coalport Brook and the enormous flow caused the excavation of the Ironbridge gorge through the Wenlock Limestone escarpment down to a level of 90 m O.D. At the same time, Lake Coalbrookdale was lowered and extended to Buildwas, and this combined lake was called Lake Buildwas.

While Lake Buildwas was forming, the retreating ice around Newport caused a lake to form against high ground to the east. This glacial lake, Lake Newport, overflowed eastwards at Gnosall towards the Trent drainage, and the overflow channel is at present at 95 m O.D.

As the ice front retreated northwards Lake Newport and Lake Buildwas eventually joined to form Lake Lapworth, but as the exit through the

Ironbridge gorge was a little lower (90 m compared with 95 m O.D. at Gnosall) the combined waters drained permanently through the Ironbridge gorge. Lake Lapworth increased further in size, as has been explained above, to cover large areas of the North Shropshire Plain. This is all Professor Wills' story, and since his classic work in the 1920s more detailed studies have suggested that the story is perhaps a little more complicated, although the basic idea of the cutting of the Ironbridge gorge by overflow of glacial lakes is accepted. Great masses of fluvio-glacial gravels and sands (Plate 31) were deposited at the entrance to the gorge near Buildwas. As the gorge grew deeper and drained Lake Lapworth, the headwaters of the Severn above Shrewsbury joined the new direction of flow and the river never returned to its old course, even after all the ice had disappeared. So we can thank the Ice Age for giving us the River Severn in Shropshire and the Ironbridge gorge (Plate 32).

Other glacial lakes were formed in Shropshire and North Herefordshire by masses of ice blocking off previous river channels, and these also led to spectacular new diversions. The Teme used to flow from Leintwardine southwards, but Welsh ice blocked its exit north of Aymestry and a glacial

Plate 32. The Ironbridge gorge formed during the melting of the Devensian ice sheet about 15,000 years ago. Note the Wenlock Limestone on Lincoln Hill (top left hand corner), the only outcrop of Wenlock Limestone north of the River Severn. Most of Ironbridge is situated on Coal Measures, and the whole area is noted for its instability, particularly around Lincoln Hill, due to underground workings in the Wenlock Limestone.

lake, Lake Wigmore, formed in the present Vale of Wigmore, and eventually overflowed northwards to cut the spectacular Downton Gorge in which the Teme now flows northwards and eastwards to Ludlow. The lake sediments of the old Lake Wigmore and the outline of its shape are obvious features in the area today.

The Upper Onny flowing down from the Shelve area and western Longmynd used to flow west into the Camlad, but its exit was also blocked westwards by Welsh ice near Church Stoke. A lake formed in the flat area between Lydham and the Longmynd which overflowed east to cut the Onny gorge between Plowden and Horderley and down towards Craven Arms. Many other river diversions were caused in Shropshire by the Ice Age.

Glacial Features of the Church Stretton Valley

The Church Stretton valley can be used as a good example to demonstrate many typical features of glaciation. It has also been studied in great detail and provides a fascinating story.

Firstly we must dispel any myths about the valley having been overdeepened by the gouging action of a moving valley glacier, in the way that has occurred in the more mountainous areas of north west Britain, causing a deep 'U' shaped profile. It has not! In fact Church Stretton was the southern limit of the Irish Sea ice in Devensian times, the ice reaching as far south as Little Stretton. A tongue of ice filled the valley up to 260 m (850 feet) O.D. There were probably two advances in this direction before the ice finally started to retreat about 15,000 years ago. To the south of Little Stretton, Welsh ice of the early Wolstonian glaciation reached as far east as Marshbrook, coming round the southern flank of the Lonymynd. Prior to the Devensian the Church Stretton valley was not as deep as it is today. It was then cut to a deeper level than today during the late Wolstonian, and finally filled up by sand, gravel and boulder clay to its present level during the late Devensian and Holocene.

A concealed valley floor now lies between 30 m (100 ft) and 60 m (200 ft) below a thick covering of gravel deposits from All Stretton in the north through Church Stretton to Little Stretton. When was this buried valley cut and what was its previous level? Prior to the Devensian the base of the valley was probably between 215 m (700 ft) and 245 m (800 ft) O.D. The broad, flat isolated top of Brockhurst is a planed rock surface at 215 m (700 ft) O.D. in the middle of the valley, and Allen Coppice close to the Longmynd Hotel is a flat surface at 245 m (800 ft) O.D. The relatively flat upper part of Ashes Hollow is probably a remnant of the stream slope flowing into this older and higher main valley. Brockhurst and Allen Coppice are probably old remnant erosion surfaces of the old (Tertiary or Anglian) valley bottom cut into by the melt waters associated with the

Wolstonian glaciation. This water erosion cut the valley down deeply to a new level between 150 m (500 ft) and 140 m (460 ft) O.D. sloping gently south from All Stretton to Little Stretton, and this new valley has a classic 'V' shaped river profile proved by a number of cross sections based on boreholes and gravity surveys. Brockhurst then became isolated from the main Longmynd and still remains so.

The Irish Sea Devensian ice then moved south into this already overdeepened valley and on its retreat left behind boulder clay, sand and gravel between 30-60 m (100-200 ft) thick to give the valley its present topographic height, 190 m (620 ft) O.D. at the watershed near to Church Stretton, and a flattish hummocky, badly drained surface. The boulder clay contains erratics from Scotland and the Lake District, particularly granites.

When the Devensian ice advanced into the valley it reached up to 260 m (850 ft) O.D. on the sides of the Longmynd between All Stretton and Church Stretton, and the Caradoc range to the east. Boulder clay occurs at 260 m (850 ft) O.D. at the top of Sandford Avenue on the road to Hope Bowdler, and so on the eastern side of the valley the ice just lapped over the lip of the confining Uriconian Volcanic barrier at Sandford Seat. Further south the ice front descended to below Brockhurst at 215 m (700 ft) O.D. since the flat planed surfaced shows no signs of having been ice moulded or modified in any way.

However, the melting of the ice between Church Stretton and All Stretton has left some remarkable glacial overflow channels at successive heights on the eastern slope of the Longmynd as the level of the ice fell. Water flowed directly eastward off the Longmynd during the retreat of the ice, but when it met the north-south trending edge of the main ice in the Church Stretton valley, it was diverted southwards along the edge of the ice sheet which sloped from north to south. This water flow between the hillside and the ice edge cut out small valleys parallel to the main valley and these can now be seen, completely dry, running in a series of steps on the eastern slope of the Longmynd, particularly between Cardingmill Valley and All Stretton, each lower level corresponding with a retreat phase of the ice. A very good example, on a small scale, is the first fairway of Church Stretton golf course, a shallow north-south channel about 150 m long. Others on a larger scale occur around Cwmdale, Novers Hill, and Castle Hill, All Stretton.

The ice had disappeared from the Church Stretton valley by 13,000 years ago, one of the final stages being the formation of a small glacial lake north of Brockhurst, now seen as flat boggy area with peat deposits, which floods quite often. At around 13,000 years ago, and onwards for a few thousand years, a harsh climate, produced high rainfall which speeded up the deep downcutting of the numerous 'batches' and valleys of the eastern Longmynd. Great floods of sand and gravel come out of these valleys and spread out into the main valley floor as alluvial fans and cones, which can be clearly seen at the entrance to Cardingmill Valley, the Batch and Ashes Hollow.

The present streams are not powerful enough to have produced these very steep sided, young, valleys with their classic interlocking spurs. The valleys are evidence of a long period of intense fluvial (water) erosion in the late Devensian and the early part of the Holocene or post-glacial epoch, nowadays called the Flandrian stage. The climate is now drier and the rapid downcutting of the batches has now ceased.

The Holocene, or post-glacial, epoch

The last 10,000 years are the first part of the Flandrian stage of the Holocene epoch. The Holocene is the second epoch of the Quaternary period, following on from the Pleistocene epoch. In Britain it is sometimes called the Post-glacial or Recent. Post-glacial it certainly is at present, but as has been said earlier, we cannot be sure that the ice sheets will not return.

During the last 10,000 years Britain has been free of ice, except for the highest mountains in Scotland, where the ice lingered longest. The climate has become warmer, but actually it was at its warmest, 1-2°C more than today, in the early Holocene, and has cooled slightly since. The rapid melting of the Devensian ice caused a rise in sea level between 13,000 and 7,500 years ago and the sea transgressed over many coastal areas in what is called the Flandrian transgression. This transgression was to a certain extent offset by the isostatic rise of the land following the melting of the ice, but the eustatic rise in sea level was greater than the isostatic readjustment, leading to the Flandrian transgression which finished 7,500 years ago, and finally cut us off from the continent. Since that time, further uplift has led to a regression of the sea and left the Flandrian shorelines some 15 m above present sea level in parts of northern Britain, shown in what we call raised beaches.

In Shropshire the rapid erosion of the hill areas by fast flowing streams, such as on the Longmynd, has ceased, and the valleys have been filling with alluvium — sand, silt and gravel laid down by large streams and the major rivers — and this is still forming today. Some of the Pleistocene kettle holes have been filled up with peat in the last 10,000 years, and peat deposits have also filled up small glacial lakes on the fringes of the ice sheets such as in the Church Stretton valley, and around Marton Pool, west of the Shelve area. Some kettle holes still retain water as in our famous meres. High level peat has also formed on some of the high ground, such as the Longmynd.

The present day alluvium of the main river valleys, such as the Severn, forms a flat flood plain over which the river periodically floods to lay down more alluvium. Meandering through its flood plain above Ironbridge, the

Severn is also downcutting through its own alluvium and redepositing it downstream.

As the Severn has been downcutting into its flood plain for 10,000 years or more in a sporadic manner, it has left behind at successive higher levels than the present alluvium, flat river terraces, the alluvium of earlier periods when the river was at a slightly higher level. These old river terraces have a patchy distribution on either side of the river but can be mapped at successive heights of a few metres above the present flood plain. Three terrace levels can be seen in the Severn and its major tributaries above Ironbridge.

These terraces occasionally contain animal bones which tell us something about the climate, and they can be correlated with terraces in other river valleys to indicate sudden changes in the rate of erosion. Height above sea level is not a definite criterion for correlation, as clearly a terrace formed 6,000 years ago in the Severn near Bridgnorth is going to be at a lower level than a terrace of the same age at Welshpool.

The formation of alluvium brings us to the end of the geological history of Shropshire — at least, it brings us up to the present day.

What of the Future?

The processes of erosion and deposition, indeed all geological processes are continuous, as I have so often stressed, and if we consider what we have learned of the past we can suggest what might happen to the British Isles in the next few million years or so.

In the more immediate future the Ice Age may return because we may now only be in an interglacial phase. In the long term the widening of the Atlantic at the expense of the Pacific Ocean may cease, and the Atlantic may start to close, with subduction zones forming on either side. A subduction zone off western Europe would produce volcanoes over Britain with very intense earthquakes! Once again, as happened in the late Precambrian, Shropshire could be covered by rhyolite lavas and ashes, and the Church Stretton Fault could be reactivated and resume its earthquake activity. If this happened we could say we could be back where we started at the beginning of this book, 650 million years ago.

Selected References

ANDERTON, R., BRIDGES, P.H., LEEDER, M.R. and SELLWOOD, B.W., *A Dynamic Stratigraphy of the British Isles*, George Allen and Unwin, London, 1979

BASSETT, M.G., et al., 'The type Wenlock Series', *Report of the Institute of Geological Science*, No. 75/13 (1975). 19pp.

BRUNSDEN, D., GARDNER, R., GOUDIE, A. and JONES, D., *Landshapes*, David and Charles, Newton Abbot, in association with Channel Four Television, 1988

COCKS, L.R.M. and FORTEY, R.A., 'Faunal evidence for oceanic separations in the Palaeozoic of Britain', *Journal of the Geological Society*, Vol. 139 (1982) 465-478

DEAN, W.T. and DINELEY, D.L., 'The Ordovician and associated Precambrian rocks of the Pontesford District, Shropshire', *Geological Magazine*, Vol. 98 (1961), 367-376.

DINELEY, D.L., 'Shropshire Geology: An Outline of the Tectonic History', *Field Studies*, Vol. 1 (1960), 86-108.

EARP, J.R., 'The geology of the south western part of Clun Forest,' *Quarterly Journal of the Geological Society of London*, Vol. 96 (1940), 1-11

EARP, J.R., 'The higher Silurian rocks of the Kerry District, Montgomeryshire', *Quarterly Journal of the Geological Society of London*, Vol. 94 (1938), 125-160

EARP, J.R., and HAINS, B.A., 1971 British Regional Geology — *The Welsh Borderland* (3rd Edit). H.M.S.O., London

HAINS, B.A. and HORTON, A. 1969. British Regional Geology *Central England*. (3rd Edit). H.M.S.O., London.

MARTINSSON, A., BASSETT, M.G. and HOLLAND, C.H., 'Ratification of Standard Chronostratigraphical Divisions and Stratotypes for the Silurian System', *Lethaia*, Vol. 14 (1981), 168

McKERROW, W.S., *The Ecology of Fossils*, Duckworth, London, 1978

McKERROW, W.S. and COCKS, L.R.M., 'Ocean, island arcs and oliostromes: the use of fossils in distinguishing sutures, terranes and environments around the Iapetus Ocean', *Journal of the Geological Society of London*, Vol. 143 (1986), 185-191

RAMPINO, R., 'Dinosaurs, comets and volcanoes', *New Scientist*, 18 February 1989

SCOFFIN, T.P., *An introduction to Carbonate Sediments and Rocks*, Blackie, London, 1987

TOGHILL, P. and CHELL, K., 'Shropshire Geology — Stratigraphic and Tectonic History', *Field Studies*, Vol. 6 (1984), 59-101

WARRINGTON, G., et al., 'A Correlation of Triassic Rocks in the British Isles', *Geological Society of London*, Special Report No. 13, 1980

WHITTARD, W.F., 'The stratigraphy of the Valentian rocks of Shropshire. The Longmynd, Shelve and Breidden outcrops', *Quarterly Journal of the Geological Society of London*, Vol. 88 (1932), 859-902

WHITTARD, W.F., (compiled by Dean, W.T.), 'An account of the Ordovician rocks of the Shelve Inlier in west Salop and part of the north Powys', *Bulletin of the British Museum (Natural History) Geology*, Vol. 33 (1) (1979), 1-69

WOODCOCK, N.H., 'The Pontesford Lineament, Welsh Borderland', *Journal of the Geological Society of London*, Vol. 141 (1984), 1001-1014

WOODCOCK, N.H., 'Strike-slip faulting along the Church Stretton Lineament, Old Radnor Inlier, Wales,' *Journal of the Geological Society of London*, Vol. 145 (1988), 925-933

WOODCOCK, N.H. and GIBBONS, S.W., 'Is the Welsh Borderland Fault System a terrane boundary?', *Journal of the Geological Society of London*, Vol. 145 (1988), 915-923

ZIEGLER, A.M., COCKS, L.R.M. and McKERROW, W.S., 'The Llandovery transgression of the Welsh Borderland', *Palaeontology*, Vol. 11 (1968), 736-782

British Geological Survey Publications

The British Geological Survey (BGS) has published, as well as their British Regional Geology series referred to above, various detailed geological maps covering parts of Shropshire, usually on a scale of 1:50000, in both Solid and Drift editions. These cover the following areas, sheet numbers in brackets: Wrexham (121), Nantwich (122), Oswestry (137), Wem (138), Stafford (139), Shrewsbury (152), Wolverhampton (153), Church Stretton (166), Dudley (167), Droitwich (182).

Explanatory memoirs have also been published to accompany each map, e.g. *The Geology of the County around Oswestry (Sheet 137)*, etc. Many of the memoirs and some of the maps are now out of print, but usually available in local libraries.

More detailed geological maps on a scale of 1:25000 are available (all in print in 1989) covering Church Stretton (Sheet SO49), Craven Arms (SO48), Wenlock Edge (SO59), Leintwardine and Ludlow (parts of SO47 and SO57), Telford (SJ60 and parts of SJ61, SJ70 and SJ71). The first three have detailed explanatory memoirs published separately. All BGS publications can be purchased from H.M.S.O. (Government Bookshops). Open file reports and unpublished BGS maps of the Shelve area are available for inspection at BGS headquarters, Keyworth, Notts, NG12 5GG.

Index

Abon Burf 134
Abdon Limestone 112, 114
Acton Reynald 160
Acton Scott 36, 81, 82; — Group, 77, 81
Alberbury Breccia (Cardeston Stone) 141-143
All Stretton 103, 177, 178
Allen Coppice 177
Alpine Orogeny 158, 160
alternata Limestone 77, 81, 96
Andesite 6, 24-26, 68, 74, 75, 85-87; amygdaloidal, 25
Anglian glaciation 163, 164, 166, 167, 169
Ape Dale 11, 55, 97, 100
Arenig Series (Ordovician) 59, 60, 62, 63, 65, 67
Ashes Hollow 177, 179
Ashgill Series (late Ordovician) 61, 62, 65, 70, 82, 83, 92
Audlem 152
Avalonia 46, 47, 59, 62, 105
Avicula contorta Beds 153
Aymestry Limestone 100-103

Baltica 44-47, 59-62
Barytes 11, 38, 65, 72, 75-77, 130, 150, 154, 155
Basalt 131, 159, 161; amygdaloidal 26
Baschurch 148
Batch Valley (Longmynd) 10, 30, 32, 178
Batch Volcanics 34, 37
Bentonite (soft clay) 37, 38, 100
Berwyn Dome 70, 83, 85, 105, 118
Berwyn Mountains 64, 92, 105
Betton Beds 70
Bishop's Castle 103, 104
Boundary Fault 125, 128, 135
Breidden Hills 11, 64, 85-87, 104-106, 175; anticline 84, 85; Ordovician rocks on, 86
Bridges Formation 34, 40
Bridgnorth 15, 16, 132, *141-144*
Brockenhurst 177, 178; — Fault, 149, 150
Brockton Fault 25
Broseley Fault 125
Brown Clee Hill 101, 110, 111, 112, 119, 134; syncline, 122, 134, 135
Buildwas Formation 97
Bulthy Formation 86
Burway Formation 34, 37
Buttingham Formation 104
Buxton Rock 34, 37

Caer Caradoc 10, 20, *25-29*, 39, 53, 55, 56, 77, 178; cross section, 27, 56
Caledonian (Acadian) Orogeny 47, 59, 61, 62, 85, 89, 91, 93, *104-106*, 111, 114, 121, 131, 134, 135

Cambrian 8, 10, 11, 17, 28, 36, 38-42, 61, 62, 65, 67, 78, 88, 119, 143; background, *43-48*; in Shropshire, *49-57*
Camptonite 51
Caradoc Series (Ordovician) 59, 60, 62, 63, 65, 66, 67, 72, *76-83*, 84, 86, 96, 104
Carboniferous 11, 15, 47, 76, 83, 85, 106, 112; background, *115- 117*; in Shropshire, *117-136*
Carboniferous Limestone 15, 83, 115, 116, *117-121*, 122, 125, 129, 141, 153
Cardingmill Grit 34, 37
Cardingmill Valley 30, 36, 37, 178, 178
Cardington 96; — Hill, 28, 77
Catherton Common 133
Causemountain Formation 104
Cefn Formation 104
Cefn-y-fedw Sandstone 121, 129, 136, 141
Charlton Hill 25, 53
Chatwall Flags 77, 81
Chatwall Sandstone 77, 81, 82
Cheney Longville Flags 77, 81
Church Stretton 1, 28, 44, 63, 72, 77, 78, 94, 177, 178; — Hills, 11, 77
Church Stretton Fault 11, *21-27*, 30, 37-40, 50, 53, 55, 63, 77, 82, 92-94, 96, 102-106, 160, 180
Church Stretton Valley 10, 16, 23, 25-27, 103; glacial features of, 170, 171, *177-179*
Clays, coarse ceramic 122, 124, 128, 131, 134, 156
Clee Group (Devonian) 113, 114, 131, 134
Clee Hills 11, 15, 96, 101, 102, 105, 106, 109, 112, 114, 135, 136, 171; — Coalfield, *132-134*
Cleobury Mortimer 96, 102, 114, 131; synclinal basin, 135
Clive 153, 154, 160
Clun Forest 11, 92, 93, 104-106; — Disturbance, 92
Coal Measure Cyclotherm 122, 123
Coal Measures 15, 38, 39, 76, 85, 100, 102, 115, 119, *122-136*, 141, 143, 145
Coalbrookdale coalfield 117, 119, 122, *125-129*, 132, 135, 136
Coalbrookdale Formation 97
Coalfields in Shropshire *124-134*
Coalport Formation 128, 129
Coed-yr-Allt Beds 128, 129
Comley 67; — Quarry, 44, 53
Comley Limestone, Lower 50, 51, 53-55
Comley Sandstone, Lower and Upper 49, 50-53, 55, 56
Comley Series (Lower Cambrian) 43, 44, 50
Condover Mammoth vii, 16, 168, *171-175*
Conglomerates 38-40, 49, 50, 67, 83, 86,

95, 112, 141, 144
Continental Drift 2, 3, 128, 165, 166
Copper 38, 75, 77, 118, 153, 154, 161
Cornbrook Sandstone 122, 132, 135
Corndon Hill 1, 2, 74; cross section 73
Cornstone/Cornstone conglomerates 112-114
Corve Dale 11, 96, 100-102, 105, 112, 134, 171
Cothercott Hill 72
Craven Arms 11, 39, 81
Cretaceous 15, 138-140, 158, 159, 161
Criggion 85, 87; — Dolerite, *85- 87*
Criggion Shales 86, 87
Culm Measures 116
Cwms (Caer Caradoc) 39, 55
Cwm Mawr 74

Deacorner Fault 125
Deltaic Sandstones 121, 122, 124
Devensian glaciation 164, 166, 167, 170, 177-179
Devonian 6, 11, 22, 38, 47, 61, 62, 76, 85, 88, 91, 93, 94, 101, 102, 105, 106, 116, 122, 131, 132, 135; background, *107-109*; in Shropshire, *109-114*
Dhustone 15, 132, 133, 134
Diachronism 48, 49
Diagenesis (in formation of dolomite) 118, 151
Dinantian Series (Carboniferous) 115, *117-121*
Disconformity 67
Ditton Platform 112
Ditton Series 105, 106, 111, 113, 114
Dolerite 15, 25, 26, 29, 40, 74, 132, *133, 134*, 160, 170
Dolgelly Beds 50
Dorrington esker 168, 171
Doseley 119
Downton (Pridoli) Series (Silurian) 91, 93, 101, 102, 104, 111
Downton Castle Sandstone 101, 104
Downton Gorge 91
Dune Sandstones (Bridgnorth Sandstone) 30, *142-144*, 150; cross bedding in, 143
Dyke dolerite 25, 26, 40, 74, 145, 154, 155, 159, 160, 161; Neptunean, 78, 79

Eardiston 153, 154
Earl's Hill 10, 28, 29, 38, 83
Earthquake activity 3-5, 9, 11, 21, 47, 61
East Shropshire Plain 141, 143
Easthope 98
Edgton Limestone 103
Enville Beds 141, 143
Erbistock Beds 130
Ercall 24, 50, 51; — Granophyre, 41, 42, 50
Esk Bed 148, 150

Farley Member (= Tickwood Beds) 63, 97, 98
Farlow Group 113, 114, 121

Faults, types and formation of 5, 9, 13, 21, 22, 23, 61, 89, 106, 149, 154, 160, 161; Church Stretton, 26-28, 31, 38, 40; Shelve, 70-73; Caradoc, 82, 83; coal measures, 125; Variscan orogeny, 134, 135
Fireclay 15, 123-125, 127, 129
Flandrian Stage (Holocene) 179
Fold Formation 9, 13, 28, 61, 70, 108, 111, 116, 131; Longmynd, 30-33, 40-42; Ordovician, 82-84; Silurian, 88, 89; Caledonian Orogeny, 105, 106; Titterstone Clee, 113; Variscan Orogeny, 134- 136; Tertiary, 159, 160
Fossils 6, 8; Cambrian, 11, 40, 41, 43, 44, 49, 51, 53-55; Ordovician, 8, 61-63, 66, 67, 67, 70, 78, 80, 81, 83, 96, 144; Silurian, 91, 95-98, 99, 101, 104, 109; Devonian, 108, 109, 111, 113; Carboniferous, 118-122, 124, 127, 142, 145; Permo-Triassic, 138, 145, 148, 153; Jurassic, 156, 157

Gaerstone 28
Gannister 123
Geological Maps, Shropshire 12, 13
Geological Time Scale 6, 7, 9
Gilberries 36
Glacial and periglacial features *167-180*; in Shropshire, 170-179
Glacial Drift 16, 84, 85, 103, 106, 152, 155, 156, 160
Glacial lakes 169; in Shropshire, *175-177*, 179
Gondwana (land) 19, 44-46, 59, 60, 138, 159; — glaciation, 165, 166
Granite 25; in Shelve mineralisation, 75, 77; — erratics, 168, 171, 178
Granophyre 18, 25, 41, 42, 50
Greywacke 37-39, 67, 116
Grinshill 15, 141, *145-150, 153- 155*, 159, 161
Grinshill Sandstone (= Helsby Sandstone) 15, *145-151*, 160, 161; jointing in, 69; mineralisation, *153, 154*

Hadley Formation 128
Hagley Shales 86
Harnage 77, 82; — Shales, 78, 81, 83
Haughmond Hill 38, 39
Hawkstone 15, 141, 145, 146, 149, 150, 161
Hazler Hill 26, 78, 79
Hazler Quarry 78
Helmeth Grit 37, 38
Helmeth Hill 26, 27
Highley Beds 131, 132
Hill End 55, 56, 82
Hill Farm Formation 86
Hoar Edge 11, 78; — Grit, 70, 78, 81, 82
Hodnet Fault 149, 150
Holocene (Post Glacial) 162, 164, *179, 180*
Hope 68; — Dale 11, 100
Hope Bowdler 39, 55, 78, 178; — Hill, 28

Hope Shales 68, 72, 74, 76
Hopesay Hill 39
Horderley 103, 104

Iapetus Ocean 19, *44-47*, *58-61*, 63, 67, 68, 91; closing of, 59, 60, 88, 89, 105, 107
Ice Ages vii, 3, 15, 16, 139, 155, 160, *162-169*; in Shropshire, *169-180*
Igneous activity 18, 25, 51, 61, 73- 75, 84, 85, 87, 116, 131, 145; causing mineralisation, 154, 160, 161
Ignimbrite (volcanic ash) 24, 37
Iron Ore 15, 117, 124
Ironbridge 11, 15, 16, 96-98; 100, 117, 125, 179; formation of Gorge, 169, 175, 176
Ironstone 100, 124, 125, 127-129, 131, 132, 134

Jointing 68, 69, 76, 77, 119, 159
Jurassic 15, 121, 138, 140, 141, 149, 151-154, *155-158*, 160, 161

Keele Beds 128, 130-132, 136, 141-144
Kenley Grit 96
Kenstone Hill 150
Kettle holes 168, 171, 179; stages in their formation, 173
Keuper Marl (Mercia Mudstones) 147, 151, 153; sequence, 152
Kidderminster Conglomerate 143-145, 150

Lan Fawr Hill 75
Laurentia 44-47, 59, 60, 62, 105
Lawley 10, 25, 26, 29, 53, 55
Lead 11, 65, 75-77, 118, 130
Ledbury Shales 101
Leebotwood Coalfield 130, 131
Lias, Lower, Middle and Upper 155-158
Lightmoor Fault 125, 128, 135
Lightspout Formation 41
Lilleshall 50, *177-119*, 143; — Hill, 24
Limestone (see also Carboniferous — Wenlock — etc.) 11, 15, 30, 49, 68, 70, 82, 92, 96, 98, 100, 107, 113, 155, 156
Lincoln Hill 98, 100
Linley (see also Pontesford- — Fault) 40; — Hill, 29
Little Hill (= Primrose Hill) 18, 20
Little Stretton 170, 177, 178
Little Wenlock 117-119, 125; — Basalt, 119
Llandeilo Series (Ordovician) 62, 64, 67, 70, 84, 86
Llandovery Series (Silurian) 62, 91, 92, 94-96, 102-104, 135
Llanvirn Series (Ordovician) 62, 64, 66, 67
Llanymynech Hill 15, 118, 141; minerals 118, 119
Long Mountain 11, 92, 93, 96; — syncline, 84, 85, 104, 105; — Siltstone Formation, 104
Longmynd vii, 1, 10, 11, 23, 28, 29-33, 38, 40, 56, 94-96, 102, 103, 130, 169-171, 177-179; syncline, 30, 32, 38; — Scarp Fault, 40, 41, 160
Longmyndian Fold Structure 31, 33, *40-42*
Longmyndian Sediments 17, 18, *29-42*, 55
Longmyndian Sequence of rocks 34, 77, 78, 83, 95, 119, 130
Ludlow 9, 11, 91, 96-98, 100, 133, 177; anticline, 101, 106, 111, 122, 135; — Bone Bed, 101, 104
Ludlow Series (Silurian) 11, 91, 95, 101, 103-105
Ludlow Shales, Lower and Upper 100-103
Lydebrook Sandstone 119
Lyth Hill 38, 39, 168, 171

Maddock's Hill 51, 52, 119
Marchamley 150
Marine Transgression 32, *47-49*, 63, 92, 94-96, 103, 115, 117, 153, 179
Marl (calcareous clay) 113, 114, 128, 130-132, 141, 151; fisheye, 156
Marshbrook 27, 38, 170, 177
Meadowtown Beds 70
Mercian Highlands 117, 136, 141
Merioneth Series (Upper Cambrian) 43, 50
Metamorphism 5, 17, 18, 42, 51
Middletown Formation 86
Midland Shelf Platform 92, 93, 100; silurian rock sequence, 94
Milkanovitch theory 162, 167
Millstone Grit 15, 81, 115, 121, 122, 129
Moel-y-golfa 87
Mudstone 30, 83, 86, 92, 98, 104, 116, 127-129, 131, 151
Myddle Quarry 145, 147, 150
Mytton Flags 67, 68, 76, 77

Namurian Sereis (Carboniferous) 115, *121*, *122*, 132, 135, 136
Nesscliffe 15, 141, 145, 146, 149, 150, 161
New Red Sandstone (Permo-Triassic) 137; rocks in Shropshire, *141-153*
Newton Brook Formation 86
Norbury 97, 103
North Shropshire Plain vii, 10, 15, 16, 39, 130, 145, 151, 170, 171, 175, 176; synclinal basin, 141, 143, 149, 160

Old Red Sandstone 11, 62, 91, 101, 102, 105, 106, *107-114*, 131, 133, 135
Old Red Sandstone continent 91, 105, 107, 108, 117
Onny Shales 77, 78
Onny Valley 77, 78, 82, 96, 170, 171
Oolitic Limestone 121, 157
Ordovician 6-9, 11, 28, 36, 38, 39, 44-47, 51, 53, 55, 57, 88, 91, 96, 103, 118; background, *58-63*; in Shropshire, *63-67*; rock sequences (Shelve) 65-69, (Caradoc) 77-82, (Breiddens) 83-87
Orogeny 5, 14, 18. See also Alpine, Caledonian, Shelveian, Variscan

Oswestry 64, 83, 92, 117-119, 121, 129, 130, 144

Paleogeography, Cambrian 45
Paleogeography, Devonian 108
Paleogeography, Llandovery 89
Paleogeography, Lower Carboniferous 116
Paleogeography, Lower Ordovician 64
Paleogeography, Upper Ordovician 65
Paleogeography, Wenlock age 93
Pangaea 135, 137, 138, 139, 140
Pattingham-Titterstone Clee Fault 135
Pentamerus Beds 96, 97, 102-104
Permian 15, 117, 130, 133-136; background, *137-140*; in Shropshire, *141-144*
Picrite 74
Pim Hill 149, 153
Pitch 38, 76, 130
Pitchford 38, 130, 141
Plaish 36
Plate Tectonics *2-6*, 19, 20, 21, 40, 44-47, 72, 82, 83, 139, 158, 162
Platyschisma helicites Beds 104
Playa lakes 151, 152
Pleistocene Epoch 15, *162-179*; stages of, 163; climatic fluctuations, 166
Pontesford 63, 76, 83; — Hill, 28, 83; — Lineament, 72, 92, 106
Pontesford Shales 83
Pontesford-Linley Fault 23, 24, 28-30, 38, 40, 49, 50, 55, 57, 63- 65, 72, 78, 82, 92
Precambrian 10, *17-42*, 43, 47, 49, 77, 78, 103, 130, 165
Prees 15, 141, 152, 153, *155-157*; — or axial Fault, 149, 160
Primrose Hill Gneisses and Schists 10, *17, 18*, 50
Proterozoic 10, 17, 43
Psammosteus Limestone 111, 112, 114
Pulverbatch 38, 39
Purple (Hughley) Shales 96, 97, 102, 104

Quartz 18, 25, 37, 49, 50, 67, 76, 81, 130, 144, 151, 154; — arenite 50
Quartzite (See also Stiperstones —, Wrekin —) 50, 67, 144
Quaternary 15, 155, 160; major events, *162, 167*; in Shropshire, *162-180*

Radio carbon dating 166, 174
Radiometric dating 8, 37, 41, 42, 76, 160, 166
Ragleth Hill 26, 27; — Tuffs, 26
Rain pit (Arenicolites) 32, 34, 37, 40, 149, 151, 152
Ratlinghope 38
Reef Limestone 11, 49, 63, 91, 97, *98-100*
Rhaetic Beds 153, 156
Rheic Ocean 45-47, 59, 60, 91, 107, 116
Rhyolite 6, 24, 26, 28, 29, 50
Ritton Castle syncline 70
River diversions during the Ice Age 176, 177

Rodney's Pillar 87
Rorrington Shales 70, 86
Roundtain Hill 75
Ruabon Marl 128-130
Rushton Schists 10, *17, 18*, 42, 50
Ruyton 141, 146, 161
Ryton Sandstone (= Helsby Sandstone) 145-150
Salt deposits 137, 141, 151, 152
San Andreas Fault 3, 5, 21, 23
Sandford Seat 178
Sandstone (See also Permo-Triassic rocks) 10, 11, 15, 30, 32, 34, 35, 37-39, 49-53, 67, 68, 70, 76, 77, 81, 82, 95, 96, 101, 111-116, 121-124, 129, 131, 132, 137, 141
Sedimentary rocks, formation of 30, 32, 68
Sedimentation 32, 49, 84, 91, 92, 102, 111
Severn, River 10, 15, 16, 98, 169, *175, 176*, 179, 180
Severn Valley Fault System 85, 106
Shales 10, 11, 15, 18, 30, 32, 34, 37, 40-42, 67, 68, 70, 71, 83-87, 92, 98, 100, 103, 115, 118, 119, 121-125, 127, 131, 134, 152, 155, 156
Sharpstones Hill 39
Sheinton (Brook) 51, 53
Shelve 10, 40, 46, *62-67*, 81-83, 85, 86, 94, 102, 103, 105, 130, 177, 179; anticline, 62, 70, 71, 73, 74; folding and faulting, 70-74; mineralisation, 75-77
Shelvian (= Taconian) Orogeny 46, 59, 61, 62, 70, 82
Shineton Shales 38, 49-57, 65, 67, 78, 119
Shrewsbury 131, 133, 142, 144, 146
Shrewsbury-Hanwood coalfield 15, 104, 130, 131
Shropshire Meres and Mosses 16, 169, 171
Sibdon Carwood 81
Sill 51; dolerite, 132-134
Siltstone 103, 104
Silurian 6, 9, 11, 23, 36, 39, 40, 44, 46, 47, 49, 53, 58, 61, 62, 66, 76, 83, 85, 118, 125; background, 88, 89; in Shropshire, *90-106*; rock sequence, 94
Soudley Quarry 79
South Shropshire Hills 10, 141, 168, 170
Snailbeach 76
Spirorbis Limestones 128, 130, 131
Spy Wood Grit 70, 81
Squilver Hill 74
St David's Series (Middle Cambrian) 43, 50
St George's Land 117, 121
Stapeley Hill 68, 71; — Volcanics, 68
Stiperstones vii, 10, 11, 38, *63, 64*, 67, 68, 70, 94-96
Stiperstones Quartzite 46, 57, *67- 71*, 77
Stone House Formation 86
Stoneyhill Fault 119
Stratigraphic Record of Shropshire 14 (Including main orogenies)

Stretton Group (Eastern Longmyndian) 34, 36-39
Stretton Shale Formation 32, 37, 38, 40, 41
Subduction zones 4-6
Sweeney Mountain 121
Symon Fault 127, 128, 136
Synalds Formation 34, 37, 41

Tea Green Marls 151-153
Telford 15, 129, 151, 171, 175
Terrane 40, 72, 74
Tertiary 15, 23, 40, 106, 133, 145, 154, 175, 177; background, *158-161*
The Batch 179
Titterstone Clee Hill 101, 112, 113, 114, *119-122*, 170; syncline, 121, 122, 135; Fault, 135, coalfield, *132-134*
Todleth Hill 75
Tornquist's Sea 46, 59, 61
Tremadoc Series (Upper Cambrian) 43, 44, 49, 50, 62, 65, 67
Trewern Brook Mudstone Member 104
Triassic 15, 39, 117, 130, 133, 135, 160; background, *137-140*; in Shropshire, *141-154*

Unconformity *34-37*; diagrammatic explanation, 35
Uniformitarianism 2
Uriconian Volcanics *17-30*; 36-42, 50, 55, 78, 83, 103, 178; formation, 21, 31; within the Longmyndian, 36-42, 50, 55, 78, 83, 103; Western Outcrop, 28, 29

Variscan (= Hercynian/Armorican) Orogeny 47,106, 116-118, 122, 124, 127, *134-136*, 141
Volcanic activity 5, 9, 10, 19-29, 61, 92, 119, 139, 159; island arc type, 4, 60, 61, 66, 86
Volcanic Lavas and Tuffs (See also andesite, basalt, ryolite) 10, 24-26, 29, 37, 50, 63, 66, 68, 74, 75, 78, 80, 84-87, 119
Wart Hill 27, 38
Waterstones (Tarporley Siltstones) 145, *147-151*
Wem 144, 151, 153, 156
Wenlock Edge vii, 1, 2, 9-11, 23, 91, 94, *96-101*, 105, 106, 111, 171, 175; cuesta, 90, 100
Wenlock Limestone 2, 11, 23, 49, 97, 98, *103*, *104*, 118, 129, 175
Wenlock Series (Silurian) 91, 92, 95, 97, 103, 104, 119, 125
Wenlock Shale 96, 97, 100, 103
Wentnor Group (Western Longmyndian) 34, 36, 38-40, 55
Weston Beds 68, 70
Weston-under-Redcastle 150
Westphalian Series (Carboniferous) 115, 122
Whitchurch 151, 152
White Ash 37
White Lias 153
Whittery Volcanics 70
Willstone Hill 28
Wilmslow/Wildmoor Sandstone Formation (= Upper Mottled Sandstone) 145, *146-150*, 153
Wixhill 150, 153
Wolstonian glaciation 163, 166, 167, 170, 177, 178
Wrekin vii, 10, 11, 15, *17-20*, 24-27, 29, 52
Wrekin Fault 25, 50, 52, 53, 125
Wrekin quatzite 41, 49-51, 55
Wrockwardine Hill 25
Wyre Forest 136, 141; — Coalfield, 131, 132

Yell Bank 82
Yorton Bank 153

Zinc 75, 77, 118, 119